CANAL HISTORIES

THE CANALS OF THE WELSH VALLEYS AND THEIR TRAMROADS (CAMLASAU'R CYMOEDD A'U DRAMFFYRDD)

by

D.D. & J.M. Gladwin

with engravings by J.K. Ebblewhite

THE OAKWOOD PRESS

ISBN 0 85361 412 1

Typeset by Gem Publishing Company, Brightwell, Wallingford, Oxfordshire.

Printed by Alpha Print (Oxon) Ltd, Witney, Oxfordshire.

First Edition 1975
New Enlarged Edition 1991

We are indebted to the following for photographs:

British Rail
British Waterways Board
Canals Officer, Neath Borough Council
City of Swansea Archives
G. Davies
L.A. (Teddy) Edwards
The Late Will Evans
David and Mary Lewis
Newport Reference Library
Pontypridd Public Library
Waterways Research Centre
Welsh Folk Museum

All other photographs are from the Authors' Collection. In addition many sources have provided illustrations but the inevitable constraints of a book's length have led us to omit many that we would have liked to have included.

Published by
The OAKWOOD PRESS
P.O.Box 122, Headington, Oxford.

Contents

	Introduction	5
Chapter One	**Major Canals**	7
Chapter Two	**Minor Canals**	51
Chapter Three	**Canal Tramroads**	55
Chapter Four	**Tramroad Relics**	77
Chapter Five	**Future Infinitive**	87
Appendix One	**The Neath Canal**	105
Appendix Two	**The Fourteen Locks Scheme**	106
Appendix Three	**Newport Gas Works**	108
Appendix Four	**Statistics**	109
	Acknowledgements	111
	References	111
	Index	112

Maps

Coal and Iron Industries surveyed 1842 8
Monmouthshire Canal, tramroads and railways; Cwm-bran 1875 27
Monmouthshire Canal and railways; Newport area 1900 31
Transport complex in Neath 1830–1930 33
Smith Canal and Swansea Canal, Tawe Valley, Swansea 33
Various canals and tramroads, collieries and ironworks, 1842 41
Blaenavon 'Big Pit' area 56
Bryn-Oer and Trevil tramroads: Trevil quarries 1870 58
Various tramroads and railways, Beaufort area 59
Hill's Tramroad, Blorenge area 61
Hill's Tramroad, tramplates and Blaenavon-Pwll-du tunnel 62
Various tramroads and railways, Varteg 1875 67
Various tramroads, tramplate sections 69
Sirhowy Tramroad and later railways 70
Garnddyrs Tramroad 74
Cwmcarn Colliery and Ynys-ddu Tramroad 75
Llanfihangel, Grosmont and Hereford Railways plus tramplates 76
Various tramroads, Blaenavon 81
Monmouthshire Canal: County Council and Fourteen Locks schemes 93, 98
Fourteen Locks, Malpas and Newport 95
Swansea Canal: Ynysmeudwy and various tramroads 100
Swansea Canal: various railways, etc. Clydach 1875–1950 102
Newport: drawbridge to Cordes Dos Works, 1930 108

Top end gate paddle mechanism and balance beam detail, Rheola No. 10 lock, Neath Canal, 1989. Reproduced from the original pattern of gearing used on the Neath & Tennant Canals, the teeth are of an unusual single draw type. The gate anchor is clearly visible as are the 'scotches' to enhance the grip of the boat-person's feet. A thoroughly workmanlike restoration.

Introduction

The story of the canals of South Wales and their tramroads is in the main one of a period of great prosperity for a few men – the owners – and relative poverty for many, their employees, whether working on the canals, tramroads or in ironworks, quarries or coalpits. But when these employers left the area, or transferred their affections to the railway, then began the decline of both the original industries and their dependent canals. Curiously, too often the projectors of industry and waterways in South Wales were English, most of whom, although there were honourable Quaker exceptions, treated the native Welsh like cattle. The uprisings in the 1820s and 1830s were riots of men and women driven to desperate acts against masters who, alienated by tongue and manner, could only resort to armed force, punitive hangings and transportation.

Against this background we have the waterways whose fates were so unalike: the Glamorganshire and Swansea, now almost totally obliterated; the Aberdare, which lies under the B4275/A4059 roads (Aberdare/Abercynon); the Kidwelly & Llanelly, converted to a railway, the Brecknock & Abergavenny, resuscitated and providing a haven of peace for pleasure cruisers, 'gongoozlers' and anglers alike. Horse-tramroads, converted to railways or not, have been closed and can only be traced by a cutting or embankment, often perched precariously on some steep hillside, or a tunnel or merely as some relic in a museum.

And what of the industry that gave life to the area? In some places – Blaenavon and Varteg for example – the countryside has returned to its peaceful origins, Hirwaun has disappeared under brash new industry, while those great nodal points, Merthyr, Newport and Cwmbran appear to be doing all they can to erase memories of the past, exchanging grey monstrosities, motorways and four-lane roads for the warmth, kindness and friendship that only a so-called 'slum' seems to generate.

Gymru, erfyniaf arnat I beido â chefnu'n llwyr ar y gorffennol er mwyn dyfodol a eill eich gadael yn orthrymedig unwaith eto.

Monmouthshire canal at top of Dock Street, prior to construction of Kingsway Road, 1933/34. The last commercial boat passed through in 1915; the canal in Newport officially closed in 1930. *Newport Reference Library*

Chapter One

Major Canals

In order to keep these histories as succinct as possible we have given the life of each waterway, at least up to the present era, in alphabetical order. It must be realised that each canal sits neatly in its own valley, surrounded and bounded by hills on each side, sharing its track with one or more rivers and a turnpike. A grim, sometimes inhospitable land, where canals were dug in defiance of nature, dug by men wielding picks and shovels, with only the crudest of instruments to assist the surveyors in picking a path. And these men who laid out the line of the canals, who negotiated the purchase of land, who bore down on recalcitrant contractors and mutinous labour, who rode or walked the lengths, were not the Brindleys and the Telfords, but were, without being lesser men, the ones on whom the glare of the spotlight and the heavy purse, rarely fell. William Kirkhouse, the Dadfords and Sheasbys – father and son in both cases – George Overton and John Hodgkinson were the men and under them were a multitude of Rees, Price, Murphy, McLean and many a hundred anonymous navvy. Hard men, often straight from their native hills, or from the 'twopenny ships' – twopence per head being the fare, steerage, from Ireland. Often ill-clad for the climate, maimed, diseased and homeless; all became hard or died. Bedevilled by the evils of the truck, careless of life or tomorrow, they dug, hacked and blasted their way through the unyielding recalcitrant granite of Wales, to make money for English shareholders, the owners of ironworks, coalmines and quarries who desired to have an easy access to the coast and thence to the rest of the kingdom.

Without these canals, trade in South Wales must have eventually stultified and died, for even as late as 1812 – in August, the best time for a horse to travel – we find: 'The weather being extremely wet, I was unable to proceed on horseback further than the Lamb and Flag, Cwmneath, on the 23rd being wetted through before I cd reach that place and was th'fore obliged to take a Chaise and 4 from thence to Brecon. Paid for Chaise hire and feed'g horses £3. 16. 0. Turnpike 8s. Driver 15s. Expences of self and Clerk & horses at Lamb and Flag, Cwmneath, & of self & Clerk at Brecon £1. 12. 0.'[1] Not only an expensive, unsatisfactory, journey but, worse, one made on the only alternative of the time – the Turnpike.

THE ABERDARE CANAL

TO THE PROPRIETORS OF THE ABERDARE CANAL

Gentlemen,

In Pursuance of your orders of the 11th July, I have surveyed and examined the Country between Aberdare Furnaces and the Glamorganshire, and am clearly of opinion that the best communication will be partly by a canal, and partly by a Rail Road, a plan and estimate of which I have prepared for your inspection. I recommend cutting a Canal from the Glamorganshire Canal, at the head of the 7th Lock above the Aqueduct over the river Taff, to a place called Ty Draw, nearly opposite

COAL AND IRON INDUSTRIES *SURVEYED 1842*

Aberdare Canal **Lower Glamorganshire Canal**

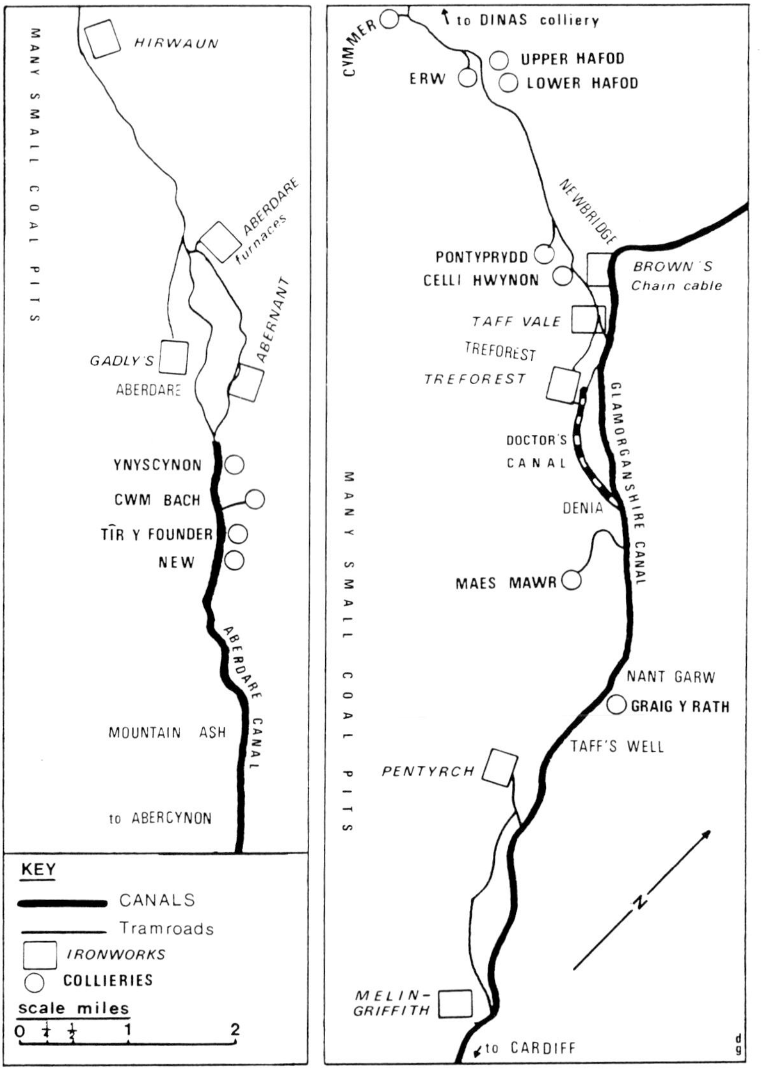

Simplified for clarity

Aberdare Village, which will be 7 miles in length, with a rise of only 25 feet, and wanting about 12 Bridges, and I estimate the same, all expences included, at ten Thousand Five Hundred Pounds.

From this Canal to the Furnaces, I propose a Tram Road one Mile and a half in length, which I estimate at One Thousand Five Hundred Pounds, including every expence.

August 1st, 1800 T. DADFORD[2]

Thus wrote the Engineer.

The canal received its Act on 28th March, 1793, with the objectives of carrying the produce of iron foundries, coal mines and limestone quarries at the following rates:

	d.	
For Iron, Timber, Goods, Wares, Merchandise, & c.	5	per ton per mile
For Iron-stone, Iron-ore, Coal, Coke, Charcoal, Bricks, Brick-tile and Slate	2	,, ,, ,, ,,
For Limestone, Building-stone, Stone, Tile, Lime, Sand, Clay and all Kinds of Manure	1½	,, ,, ,, ,,
For Cattle, Sheep, Swine and other Beasts	5	,, ,, ,, ,,
Fractions to be taken as for a Quarter of a Ton, and as for a Quarter of a Mile		

TRAVELLING ON THE RAILWAYS

	d.	
For every Horse, Mule or Ass	1	per Day
For Cows and all other Cattle	½	,, ,,
For Sheep, Swine and Calves	5	per Score[3]

On the same date as Dadford made his report, John Scale of the Aberdare Iron Company wrote:

> The Aberdare Iron Company propose, that if the Aberdare Canal Company will open a communication from the Glamorganshire Canal to their Furnaces, by a Canal as far as the same can conveniently be made, and then the remainder by a Tram-Road, and will engage to reduce the Tonnage upon the iron from their Works to three-pence per Ton per Mile, the said Aberdare Iron Company will engage, that if the Tonnage does [not] amount to the sum of Seven Hundred and Fifty pounds per Annum, to Guarantee the deficiency from the time the said Canal and Road are completed, 'till the expiration of the said Aberdare Iron Company's Lease.[2]

Prior to this the canal company had built their tramroad (*see Chapter Three*) but done little about the canal, and it was not until 1809 that Edward Martin resurveyed the line. The Committee, including the Wilkinses of Brecon, the Scales of Aberdare, Tappenden of Abernant and Joseph Bailey, accepted Thos Sheasby, Jnr, as Engineer, who resigned in 1811 being replaced by George Overton.

The canal being only 6¾ miles long and with only three locks, completion was not long delayed, it being opened for traffic in August 1812, at a cost of £26,220. For some years following the closure of Hirwaun Iron Works trade was poor and it was not until 1822, with William Crawshay II in command, that new life was injected into the waterway, when the banks were raised to

Glanusk Park near Llangattock, is one of the very attractive views to be seen on the Brecon & Abergavenny canal with the road some 70 ft below the canal and an excellent view of Sugar Loaf mountain can be obtained. The house was built in 1826 and the park laid out a couple of years later for one of the Bailey family of ironmasters; two views of the canalside stable block.

allow the passage of boats carrying 25 tons in lieu of the 15 possible hitherto, a move reflected in the dividends which rose from 1 per cent in 1814 to 5½ per cent by 1843. Faced with railway competition from 1845 onwards, dividends dropped, although, briefly, they rose in the late 1850s, reflecting local prosperity. Failing to sell themselves to railway interests and suffering a hopeless imbalance between income and expenditure, the waterway finally closed down in November 1900, its bed being sold to the Urban District Councils of Aberdare and Mountain Ash in 1923: it now lies underneath the A4059/B4275 roads.

THE BRECKNOCK & ABERGAVENNY CANAL

The first Act for the Brecknock & Abergavenny was passed by the Houses of Parliament in 1792 and shows the intention of running from Brecknock to the River Usk at Newbridge. However, anxious to obtain the traffics which it was presumed would flow along this line . . .

> an agreement being entered into between the proprietors and those of the Monmouthshire Canal, the latter paying the former three thousand pounds, it was determined to form a junction between the canals at Pont-y-Moel, below Pont-y-Pool, a fresh Act . . . being obtained in 1793. The first portion made navigable was from the coal-mines and lime-works in the vicinity of the Clydach to Llangynider bridge, a distance of eight miles and a half; and from the latter place the canal was continued ten miles further, to Brecknock, and was completed in 1801, forming an important line of carriage for coal and lime to that town.[4]

The main objective in building the canal was stated to be the entirely laudable one of 'lowering the price of coals, lime, etc.' In this it was entirely successful, for when the first portion was completed in December it had the immediate effect of cutting the cost of coal in Brecon from 12*d*. (5p) or 14*d*. (6p) per hundredweight to 9*d*. (3½p). Lime, extensively used in agriculture, was almost halved in cost. But even so this was not an easy canal to build, for the red sandstone, fractured in many places and porous by its nature, required much puddled clay to make it watertight. Cattle by the score were hired to trample the clay into a sticky mess; when they were not available men in big boots were used, as they are today.

Naturally enough these engineering works were reflected in the capital of the company. The proprietors – The Duke of Beaufort, Sir Charles Morgan and Richard Hill among them – were allowed a capital of £100,000 with an additional £50,000 if necessary. By 1800, with the canal only open as far as Gilwern, £129,000 had been spent, be it not all on the canal, as about 9 miles of tramroad had also been finished; and in 1802, probably heartened by the payment of their first dividend, it was decided to extend the canal from Gilwern to Llanfoist, the nearest point reached to Abergavenny. Also around this period Thomas Dadford, who had been Engineer of the waterway since 1795, was replaced by Thomas Cartwright.

Originally considered a sound speculation, in 1800 a commentator reported on the Brecon & Abergavenny Canal: ' . . . the enormous expence of carrying it through a mountainous district, in which the excavations must be made to a great depth, renders it uncertain whether it will ever reach the

Brecon & Abergavenny canal. In the late 19th and early 20th centuries it was common practice for a normal working canal boat to be scrubbed out, clean cloths laid in the hull, and school benches borrowed to provide an economic and pleasant outing. One hopes the crates of beer on the fore-end of this Brecon & Abergavenny boat are for the use of the crew; *c.*1912. *Welsh Folk Museum*

place of its original destination.'[5] Some of the pessimism may have been brought about by a report from Benjamin Outram in 1799.

> I have examined the line of the proposed canal between Clydach and Pont-y-Pool, and observe it to be on the same kind of open strata as that on the northern part of the line, except near the south end, where the soil appears of a better kind for holding water, and am of the opinion the execution of this 14¼ miles of canal, cannot be completely perfected for a less sum than £44,000 exclusive of the purchase of land and temporary damages, which I suppose will amount to £6,000 making together £50,000.
>
> A rail-way of the best construction would be laid along this line for £19,000 exclusive of the purchase of land and damages, which would not be half as much as for a canal; but rating it as half, the whole expense of the rail-way would not exceed £22,000. As to the comparative expenses of conveyance on canal and rail-way, they must be considered as follows: a canal on the part of the line in question would be on a perfect level, unobstructed by locks, and of course as favourable as possible for navigation. A rail-way on a perfect level is more expensive to work, than one that has a gentle fall in the direction of its greatest trade; but on a rail-way perfectly level, with waggons of the best construction, three horse managed by one driver, will draw 15 tons at the rate of 3½ miles per hour. The tonnages to the Company would be the same on a rail-way as on a canal. The freight on a canal, including wear and repair of boats, would cost more than 1*d*. per ton per mile. The expense of gauging, including wear and repair of waggons, would not amount to 1⅓*d*. per ton per mile, so that in the question before us, the canal would have the advantage of about one farthing per ton per mile, or 4*d*. per ton on the whole length between Clydach and Pont y Pool; and on articles passing between Brecon and Newport, the advantage in favour of the canal would amount to 6*d*. per ton, on account of the re-loading on the rail-way plan, it being already canal between Clydach and Brecon. On the other hand, coals, lime or other articles to be delivered at the Goedtref, or any part of the line would be conveyed full as cheap by rail-way as by canal, because by continuation on rail-way, the expenses of shipping and unshipping would be avoided . . . [6]

By 1805, by spending another £9,500, the canal managed to reach Blaen-avon Road, just short of Llanfoist, and stopped there until 1809, when Richard Crawshay lent £30,000, and other interested bodies sufficient to make a total of £50,000. Engineers again changed, with William Crossley taking over, and finally on 7th February, 1812,

> The Committee set out to view the Canal towards Pontymoile and at Mr Waddington's Boat House in Llanover, met him, and embarked there on board his Boat, and proceeded in the same all the way from thence to, and through, the stop Lock at the Junction with the Monmouthshire Canal, being the first entrance from this Canal into that, amidst the acclamation of a very numerous body of the inhabitants as a token of their Joy at an Event so very beneficial to this Country.[7]

Aside from benefits to the proprietors which this canal brought, its side effects upon the population are very clear, for example of Aberbaidon it was said that the lime, coal and Clydach ironworks: ' . . . afford employment to a large proportion of the inhabitants and the produce of which is distributed, by means of the canal, throughout the adjacent district.'[4] and for Llanvigan, Maesgwartha and Y Prysg it was said that their coal mines, limestone

ANNO TRICESIMO TERTIO

Georgii III. Regis.

C A P. XCVI.

An Act for making and maintaining a Navigable Canal from the Town of *Brecknock* to the *Monmouthſhire* Canal, near the Town of *Pontypool*, in the County of *Monmouth*; and for making and maintaining Rail Ways and Stone Roads from ſuch Canal to ſeveral Iron Works and Mines in the Counties of *Brecknock* and *Monmouth*.

WHEREAS the making and maintaining of a Canal for the Navigation of Boats, Barges, and other Veſſels, from the Town of Brecknock into the Monmouthſhire Canal, at or near a Place called Pontymoile, near the Town of Pontypool, in the County of Monmouth, will open an eaſy Communication between the County of Brecknock, the Town of Abergavenny, and other interior Parts of the County of Monmouth, and the ſeveral Ports and Navigations of this Kingdom, and the making and maintaining of Rail Ways, or Stone Roads, for the Paſſage Preamble

Part of the preamble to the official notice of the first Act for the building of the Brecknock [Brecon] & Abergavenny Canal, 33 Geo. III Cap. 96 (Royal Assent 28th March 1793).

The Glamorganshire canal staircase locks at Pontypridd *c*.1900. Opened for navigation by 1792, repairs were needed to this length even before the whole waterway was opened two years later. Disused around 1900, these locks were abandoned in 1915. *Pontypridd Public Library*

The products of Nantgarw pottery are now sought after. A typical Glamorganshire boat lies alongside the pottery wharf *c*.1890. Many boatmen lived in the village, and since these were officially only day boats, it is said 20 or 30 might lie overnight.

quarries and lime kilns were vastly expanded as the goods could easily be conveyed by the Brecknock Canal. The effect upon its terminal town, Brecon [Brecknock] was no less marked:

> The trade is principally in wool, leather, and hops, and the supply of the neighbourhood with various articles of consumption, . . . has been greatly increased since the construction of the Brecknock and Abergavenny canal [which] communicates with the Monmouthshire Canal, and thence, by Newport, with Bristol and other parts of the kingdom. The whole extent of this canal is forty-five miles, the greater portion of which is a fine level; within a few miles from the town it passes through a tunnel, nearly quarter of a mile in length, which has been cut through a hill; and on its near approach to Newport, the navigation is partially impeded by the numerous locks, rendered necessary from the inequality of the ground through which it passes. Regular trading boats have been established, which pass weekly between this place and Bristol, and capacious wharfs for coal and lime have been constructed on the banks of the canal.[3]

The tunnel and the plethora of locks were not the only engineering features, for the 'embanked aqueduct', 84 ft above the bed of the river Usk still stands in a good state of repair at Govilon. Although in the beginning the Monmouthshire Canal was most willing to help the Brecon & Abergavenny, this state of affairs was not to continue. Relying upon tramroads as main traffic arteries the company was vulnerable to sniping at the sides, and when the Monmouthshire extended its tramroad from Aberbeeg to Nant-y-Glo, concurrently reducing tolls, the Brecon & Abergavenny had no choice but to cut its own rate from 2*d.* (1p) per ton in 1828 to 1*d.* (½p) in 1830, which was reflected in their dividend of £9 in 1828, £4 in 1832. Crying 'enough', the shareholders stated ' . . . it would be highly beneficial to the Interests of this Concern that a satisfactory arrangement should be entered into with the Monmouthshire Canal Company respecting the Rates of Tonnage to be charged on the two Canals & Roads on Goods and Merchandise conveyed from the one along the other whereby the injurious competitions which have hitherto existed may be removed . . .',[8] and some talk of amalgamation took place although the marriage of these two navigations did not take place until September 1865 after many flirtations with various railway companies and after trade had fallen to a point where profits had turned to losses, the Monmouthshire concern buying the Brecon & Abergavenny for roughly £61,000.

THE GLAMORGANSHIRE CANAL

February 1794: 'The canal . . . is completed and a fleet of canal boats have arrived at Cardiff . . . to the great joy of the whole town. This canal is 25 miles long, it passes along the sides of stupendous mountains.'[9] But by 1904, ' . . . owing to many serious subsidences caused by colliery workings, the section of the canal between Merthyr Tydfil and Abercynon has been closed since 6th December, 1898, and up to the present time has not been reopened. The canal between Cyfarthfa and Merthyr Tydfil Warehouses has

The Glamorganshire canal had locks measuring 60 ft long x 8 ft 9 in. wide. This view shows North Road lock No. 49 Cardiff, *c.*1910. The gas light adjacent to the lock was an unusual provision required by the landowner.

Lock Lewis, Rhydyfelin, near Denia on the Glamorganshire canal around the turn of the century. The open windows in the building seem a little odd in mid-winter.
Pontypridd Public Library

Two further views of the Glamorganshire canal with Cardiff Castle (*top*) in winter time *c.*1920s and the Castle Lodge (*bottom*) with the castle just visible on the left. Both the paddle gear and a keb or canal rake are clearly visible in the lower photograph.
Welsh Folk Museum

been disused since 1865.'[10] This briefly sums up the history of one of the most successful – or at least, profitable – of Welsh Valley Canals, the Glamorganshire.

Built at the behest of the great ironmasters of Cyfarthfa, Merthyr Tydvil, Penydarren and Dowlais, the first survey was carried out in March 1790, the Act following on 9th June, 1790, being for the 'making and maintaining a navigable Canal from Merthyr Tidvile, to and through a place called The Bank, near the town of Cardiff, in the county of Glamorgan'[3] and including various clauses. One stated that 'Proprietors of any mines lying within four miles of any part of this canal, may make collateral cuts or railways across the grounds of other persons, on payment of damages',[3] whilst landowners could build wharves and charge accordingly. A second Act after the opening was for an extension of the canal at the seaward end to 'The Lower Layer'. The Act dated 26th April, 1796 gave two years for completion, and the Marquis and Marchioness of Bute – who were owners of the land – were to have free use of the canal and towing path below the south gate of Cardiff.

The contractors, The Dadfords and Thomas Sheasby, were without the superintendence of an Engineer, but started work in August, and by the middle of 1792 it was almost complete, although at the opening (1794) £80,000 had been spent in lieu of the contract price of £48,288. After completion of the extension (1798) the cost rose to about £103,000, giving a canal 25½ miles long, with 51 locks rising 543 feet, a short (115 yard) tunnel, and a 'stupendous and inspiring' aqueduct over the Taff at Abercynon. Problems arose, with the ironmasters at loggerheads both amongst themselves, and with the Engineers, Dadford Senior and Sheasby being arrested as they failed to make good a breach in the canal bank which occurred in December 1794. Upon arbitration by Robert Whitworth, far from getting back the £17,000 that the company thought the contractors owed them, only a paltry £1,512 was awarded.

The extension to 'The Lower Layer' was met with great acclaim: '. . . a naval procession took place, in celebration of the opening of the canal to the sea, attended by firing of guns, and the ships in the harbour, belonging to the few nations at peace with England, entering the bason with their respective colours flying'.[6]

In 1804 the Duke of Rutland made a tour of South Wales and reported upon the joys of seeing ironworks and canal both beautiful:

> We dined heartidly, and at dusk in the evening, the rain ceasing for half an hour, walked towards the forges. We wandered about for some time, and then went immediately to them, guided by the streams of fire which were bursting forth from the chimneys. The distance of them from the town is about three-quarters of a mile, and the railroad along which we walked ran by the side of a canal; as we approached them, the effect was grand and sublime beyond all description . . . The fires from the furnaces were bursting forth in the darkness of the night, and every moment we saw, as it appeared, a red-hot bar of iron walking towards us, the man who carried it not being visible. In the perspective, we could see numbers of Vulcans dragging about pigs of iron just taken from the furnaces (the fires of which would dazzle the strongest eye) and persuing their different operations, while their grimy figures, and gloomy visages, were visible by the light of the forges . . . I never saw anything that gave me more the idea of the infernal regions.[11]

Throughout a long period of the canal's history the main battle was waged between ironmasters and the canal company (whose Directors were from rival companies) on the question of the supply of water. One move by Hill of the Plymouth Works resulted in a reservoir being built at Glyndyrys in 1806 after the canal company had a report by Rennie, followed in 1809 by a 'fire engine' at Pontyrun, which lifted water from the Taff to the canal after its use in Hill's works.

Incidentally, not all the negotiating took place initially in the courtroom. On 1st August, 1794, Mr Richard Hill, Jnr a youth of 20, beat up the lock-keeper at No. 3 lock, the 'Parliament' (so called as it was here a copy of the authorising act was kept for public inspection), when the keeper tried to stop Hill from changing the water-flow away from the canal works. Eventually, on recourse to the law, William Williams, the lock-keeper, received very heavy damages – he being innocent as he was only a servant of the canal company.

Similarly problems arose over the Melingriffith Works, and in a letter dated 19th January, 1803, Richard Blakemore wrote to his solicitors that:

> . . . it has always appeared to me that we have a much broader and clearer ground of action than any that can be founded upon a breach of this particular clause, which everybody condemns as insufficient and loose, vizt., upon the ground that the canal is not executed 'doing as little damage as may be', which the canal company are bound to do by the words in the latter part of the Clause, and which we can prove is not done, inasmuch as they may and ought to supply themselves with Water without interfering with our Works, by taking the Water below the Works instead of above them.[2]

And again, another point of friction:

> There are two weirs erected near the Quaker's Yard, across the River Taff (the Canal Company have a power to divert Brooks and Streams of Water, but not the main River) whereby all the Water of that River is at some seasons of the year diverted into the Canal.[2]

In 1810 matters were obviously little better, for another letter signed by Blakemore has it that 'Our old friends the Glamorganshire Canal, are again repeating their injuries to us at Melin Griffith, which has compelled us to commence proceedings at law against them',[2] although a pump and water-wheel (still extant) had been installed by the canal company at Melingriffith.

In 1821 a new reservoir was built below the Treble Locks, Nantgarw (Nos. 35–37) to try to improve, if not resolve this matter, but even in 1835 Blakemore was still struggling, being 'extremely impatient to have the costs from the Glamorganshire Canal Company paid, and must urge you to compel the same being done without further delay.'[2]

The canal company, being bound to pay not more than an 8 per cent dividend had, for many years, an embarrassment of riches, and therefore the Directors took the unusual step of both reducing tolls and giving toll-free periods, albeit at times which suited their coal, iron and limestone demands. Given a free period they would move all they could, engaging all the boats in order to do so, which was unfortunate for their rivals.

The biggest bottleneck on the canal, aside from the number of locks, which were overcome to some extent by Sunday and night working, was the sea-port of Cardiff. In 1824 a scheme for a new dock foundered on the rock-hard opposition of the Marquis of Bute, who in turn brought out a scheme of his own and one which led the Hills and Guests, after resigning from the canal company in June 1835, to hold a meeting in Merthyr to form the Taff Vale Railway Company, the main resolution being that '. . . the present means of communication does not afford the requisite facilities and it is expedient to establish a communication by means of a Railway which shall combine the advantages of the latest improvements in the mode of transport.'[12]

In 1832 William Crawshay I of Cyfarthfa, always a realist and something of a clairvoyant wrote: ' . . . I believe railways are better than canals and will supersede them . . .'.

In 1839 the West Bute Dock was opened, giving accommodation for 250 craft within a surface area of 92,400 square yards. The increase in facilities was not noticeably reflected in the tonnage of iron moved down the canal (see Appendix 4), which 39,497 in 1817 had risen to 84,946 by 1827 and 124,810 in 1837, Dowlais, Cyfarthfa and Hirwaun representing about 60 per cent of this.

The problem by then was not to get more traffic but coping with the existing loads, a problem aggravated by the necessity to tranship from the canal to tramroads and vice versa. It was said in 1837 that 'upwards of 10,000 tons' of various goods were waiting movement at Cardiff and 'many hundred' at Merthyr* – an intolerable situation for shippers and recipients alike.

It is not surprising that the plans for the Taff Vale Railway met with little opposition, save from the canal company, their Act being obtained in 1836. From Cardiff to the Navigation House, Abercynon, the railway opened on October 1840 and six months later through to Merthyr. This led to some schizophrenia among the canal Directors, some of whom thought to blackmail such of their customers who were ill-sited for rail access by raising rates while others including William Crawshay II, especially after the failure of the tripartite (Canal, Lord Bute and TVR) amalgamation talks in 1844, thought to retain customers by a rate-cutting war.

Both parties tried their own way, the end result being the same, as from 1860 onwards tonnages steadily dropped, despite some modernisation including the building of an hydraulic coal hoist and various improvements to the locks and extensive dredging. In 1871 the general traffic level between Pontypridd and Merthyr stood at 284,041 tons, but by 1897 it was a meagre 31,113 tons. This was reflected in the dividend; gone were the bonuses previously enjoyed and the constant 8 per cent dividend, hitherto unshakeable, was halved by 1875 and had dwindled to 1½ per cent in 1882, new traffics like coal dust briquettes far from compensating for the extinction of the ironworks.

In 1883 the canal company was pleased to agree to sell all its shares to Lord Bute, this taking effect in 1885, simultaneously with his purchase of

* A similar set of circumstances was applicable on the Bridgewater Canal in England, leading to the building of the Liverpool & Manchester Railway. See *THE CANALS OF BRITAIN* (Batsford, 1973).

the Aberdare, and shortly afterward some repairs and improvements were made at a cost of about £25,000, although to little avail; the whole Bute consortium – docks, canals and tramroads – becoming part of the Cardiff Railway Company. In 1898 the section from Abercynon to Merthyr was closed and and by 1906 a writer found:

> . . . to my surprise, it had fallen into complete disuse, not a vestige of a canal boat was to be seen, and the water supply turned off into the river! Very little more than one hundred years have been long enough to bring about this marvellous change from high dividends and great prosperity to absolute ruin . . . and on the day of my visit the water in the lock was green and stagnant, and the massive lock gates had not apparently turned upon their hinges for months past![2]

In 1910, notwithstanding an attempt to use steam power in 1893, which failed due to the hindrance of the locks, 'the canal is practically derelict above Pontypridd, but the owners intend keeping it open as far as Tongwynlais to serve works en route from Cardiff', and in 1915 a breach caused the final closure of the canal above Pontypridd. As a wistful dream, a report appeared in *The Merthyr Express* of 15th January, 1956 of an interesting talk by Mr Frank Thackeray, MBE, on his 1917 campaign to restore the canal from Merthyr Tydvil to Cardiff. His plans were based on the takeover by the Government under a wartime measure applicable to all waterways, these being placed in the hands of the Canal Control Commission who were to '. . . enquire into the possibilities of reworking some of them, the Canal from Merthyr to Cardiff being included in their chart'. Just imagine as Mr Thackeray did, 'Merthyr could have had the next best thing to a seaboard; a modernised waterway for small barges floating to and from the docks at Cardiff'.[13]

The end came with another breach on 25th May, 1942, at Nantgarw, this marking the passage of the last boat; the residue of the waterway was purchased by Cardiff Corporation in 1943.

THE GLAN-Y-WERN CANAL

The Glan-y-Wern is indeed a many-faced waterway, being known as the Crymlyn & Red Jacket, Britton or Llanwern. However the facts are simple. It was made without an Act of Parliament, 'as the principal object of the proprietor [Lord Vernon] was the improvement and drainage of his own estate'.[3] The cost of building was met by the lessee of the land, Edward Elton. Completed in 1790, 3¼ miles in length from Glan-y-Wern to Trowman's Hole, nevertheless it was a prodigious feat of engineering as it was 'found necessary to take the canal for nearly two miles through the midst of Crymlyn, or Crumlin, bog or morass, the soft spongy ground of which rising up repeatedly after the surface was cut away, seemed to present an insuperable obstacle to the completion of the undertaking. The work . . . may be looked upon as not the least striking instance which this country affords of spirit and perseverance successfully exerted . . .'[1]. Elton only saw 20 years of trade, dying, probably bankrupt, in 1810, and the papers of the

Near Pontymoile (Pontymoel) Basin at the junction of the Monmouthshire and Brecon & Abergavenny canals. Junction cottage was built in 1914 at the cost of £150 although subsequently altered. The lock now minus its gates was for gauging the tolls due and for water control purposes rather than for hill climbing. At one time this cottage housed two families but it became vacant in 1979.

Brecon & Abergavenny canal. Near Pontymoile (Pontymoel) is the aqueduct over the Afon Lwyd. A single arch, it is roughly 15 ft wide and 15 ft high from the bottom of the river and, although relatively calm here in spring 1989, in winter the Afon Lwyd can become a raging torrent. So much so that the Dadford designed work of 1800 was undermined and had to be demolished and rebuilt in 1812.

solicitors employed by Lord Vernon show that a 'Notice of Distress' [writ] was served on the barges in the canal, the rent for 'the bog' being somewhat behindhand, as was that of the wharves on the Neath River at 'Trowman's Hole'; the latter place was the basin where river trows could lay up.

If we are to believe Elizabeth Davies, the canal was unlikely to bring luck. In a letter to George Tennant, dated 1st February, 1825, she claims her father, Richard Jenkins, actually proposed the waterway, having been forced to do so by the building of a turnpike gate at 'Gwindy Public House' [*sic*], the turnpike keepers having 'for 20 carts [of coal, roughly 100 tons] one pound [sterling] every day'. Jenkins had, in 1788, sent 'imediately to gloaster for an enginare, Mr Talford came down to begin the canall . . .' and arranged a lease with Lord Vernon, dying the day it was signed, 14th August, 1788. The widow then transferred the lease, and sold some land she had, to 'Squire' Elton for one hundred pounds.

Derelict or not, in 1817 George Tennant could see a future for the waterway and leased it, plus other lands from the Earl of Jersey (inheritor of Lord Vernon's estates), and was building an extended version, namely the Red Jacket, or Neath & Swansea Junction, Canal. This ran from a junction with the Glan-y-Wern about 2 miles from Red Jacket to Swansea Harbour, a total length of 4 miles, plus 1⅜ miles along what was now the Glan-y-Wern branch, the Engineer being William Kirkhouse. The logical continuation to this was the Tennant Canal (q.v.).

THE MONMOUTHSHIRE CANAL

This canal was a rare example of a two-pronged waterway. The main line ran from Pontnewyndd to Pontymoile, where it made a junction with the Brecon & Abergavenny Canal, and thence to Newport, the Crumlin branch leaving it some 1½ miles above the latter, the original Act being dated 3rd June, 1792.

The original subscribers with a capital of £120,000 included Josiah Wedgwood and the ironmasters the Duke of Beaufort, Hill (of Plymouth) and Harford & Co. (Ebbw Vale & Sirhowy). Thomas Dadford Jnr was Engineer. The main line was not navigable until February 1796, a further £60,000 being obtained meanwhile. In 1797 the company obtained a further Act enabling them to extend the waterway another 1½ miles nearer the 'Haven of Newport'. The whole waterway was finally opened in April 1799 as an 'extensive and useful undertaking, being in the very centre of a country abounding in limestone, stone, coal, iron, tin and lead, which before this work was executed were permitted to remain undisturbed, for want of a conveyance for the produce of the mines. The cost of the work has been great, but it is a speculation which will not fail eventually of proving highly beneficial to the parties who have embarked their property in it . . .'.[3]

In September of the same year the 12 months traffic figures for the whole 22 miles of waterway were published including some rather unusual items:

AMOUNT OF TONNAGE OF THE MONMOUTHSHIRE CANAL,
for One Year Commencing 9th September, 1789[5]

	Coal		*Pig Iron*		*Bar Iron*		*H. Blooms*	
	tons	*cwt.*	*tons*	*cwt.*	*tons*	*cwt.*	*tons*	*cwt.*
September Quarter	5,177	0	2,497	10	23	0	135	10
December	7,399	0	2,843	10	–	–	145	0
March	7,013	0	2,821	15	4	0	132	0
June	8,502	0	2,996	10	5	10	161	0
	28,091	0	11,159	5	32	10	573	10

	Timber		*Lime*		*Sundries*	
	tons	*cwt.*	*tons*	*cwt.*	*tons*	*cwt.*
September Quarter	82	1	114	15	354	0
December	114	0	35	10	812	5
March	50	0	–	–	354	5
June	42	5	3	0	227	10
	288	6	153	5	1,748	0

	Ale & Porter		*Stones*		*Castings*		*Bark*	
	tons	*cwt.*	*tons*	*cwt.*	*tons*	*cwt.*	*tons*	*cwt.*
September Quarter	11	10	4,777	0	45	10	26	10
December	9	0	1,928	15	38	0	0	5
March	–	–	1,130	10	45	10	–	–
June	–	–	4,517	0	7	0	36	0
	20	10	12,353	5	136	0	62	15

	Iron Ore		*Slate*		*Manure*	
	tons	*cwt.*	*tons*	*cwt.*	*tons*	*cwt.*
September Quarter	237	10	24	15	94	10
December	196	6	43	5	55	0
March	643	0	–	–	48	0
June	879	0	13	10	28	0
	1,955	16	81	10	225	10

By contrast with these dour facts a visitor in 1801 enthused over the scenery:

> From Crumlin Bridge, I continued along the side of the canal to Risca and was as much pleased with the singularity of the views in this route, as I was affected with the romantic scenery in the sequestered vale I had just quitted. The road over which I passed is the towing path, carried along a narrow ridge on the edge of the

> canal, and at the top of an elevated embankment. On the left the canal winds at the foot of overhanging rocks, fringed with wood; below the Ebwy is seen from the elevation of forty or fifty feet . . . the contrast between the tranquil waters of the canal, and the rapid eddies of the turrent, bursting through groves of alder and foaming over fragments of rock, is peculiarly striking . . . Between Newbridge and Risca, numerous houses, mills, forges and hamlets, are scattered on the side of the canal, and on the banks of the Ebwy, and the whole valley becomes a scene of bustle and activity. I passed on the left no less than four of those wooden glens which are frequent in these parts; they are called Cwms by the natives, and give a pleasing diversity to the uniform range of hills; rapid torrents rush down their hollows, and after supplying the canal, pour their superfluous waters in the Ebwy . . . .[5]

While another, an Engineer, chose instead to note: 'In the environs of the Monmouth canal, a great number of manufactories are established for the working of metals. Near a large copper foundery, is seen a cast-iron slide, constructed for the purpose of rolling down the chalk, dug out of a pit, on the other side of the canal, from the towing-path to the bottom of an adjacent meadow. This slide, which is 164 feet in length, and sixteen inches in breadth, is composed of plates about eight feet and a half in length, and fixed by strong bolts on a perfectly level ground.'[14] A common enough design, but one rarely noticed by writers interested solely in beauty.

The intended main traffic on the waterway was that of coal – black diamonds – and an astute move by the company's solicitors in the 1797 Act ensured them exemption from the export coal duties payable, at least eastward, giving them an advantage of 5*s.* 4*d.* (27p) per ton over their rivals in West Wales at a time when coal was sold for 38*s.* (£1.90) per ton delivered.

Details of the tramroads which were so vital to the existence of this waterway will be found in the relevant portions of this book; the most vital was the Sirhowy and its connections, which not surprisingly had its authority obtained within the Monmouth Canal Company's Act (26th June, 1802), the proprietors including the ironmasters Samuel Homfray, Richard Fothergill, William Thompson and William Foreman, the 'intended lessees of the Tredegar Iron Works'. The first moves were made in an agreement, dated 18th December, 1800, between the Canal and Tredegar Ironworks Companies which included clauses to the effect that:

1. The ironworks would drop the idea that they had in mind of a tramroad to be built parallel with the canal as '. . . the Canal Company were apprehensive that such tramroad might prove injurious to their interest, and being desirous also to secure the tonnage on their Canal of all iron and coal to be exported from the said New Works.'[2]

2. The Canal Company '. . . shall. on or before the 25th December, 1801, make a good and sufficient tramroad, according to the plans of Benjamin Outram, of Butterly, in the County of Derby, Engineer, from the Tredegar Iron Works, to join their Canal near Risca Church, with all convenient turnouts, and, if required, to construct a double rail.'[2]

They agreed through tonnage rates from the ironworks to the wharf in Newport at 3*s.* (15p) per ton for iron, 2*s.* (10p) for coal, the distance being reckoned as 24 miles. Coal was carried at the rate of 21 cwt. of 120 lb. to the

INDUSTRY IN THE CWM-BRAN DISTRICT, 1875

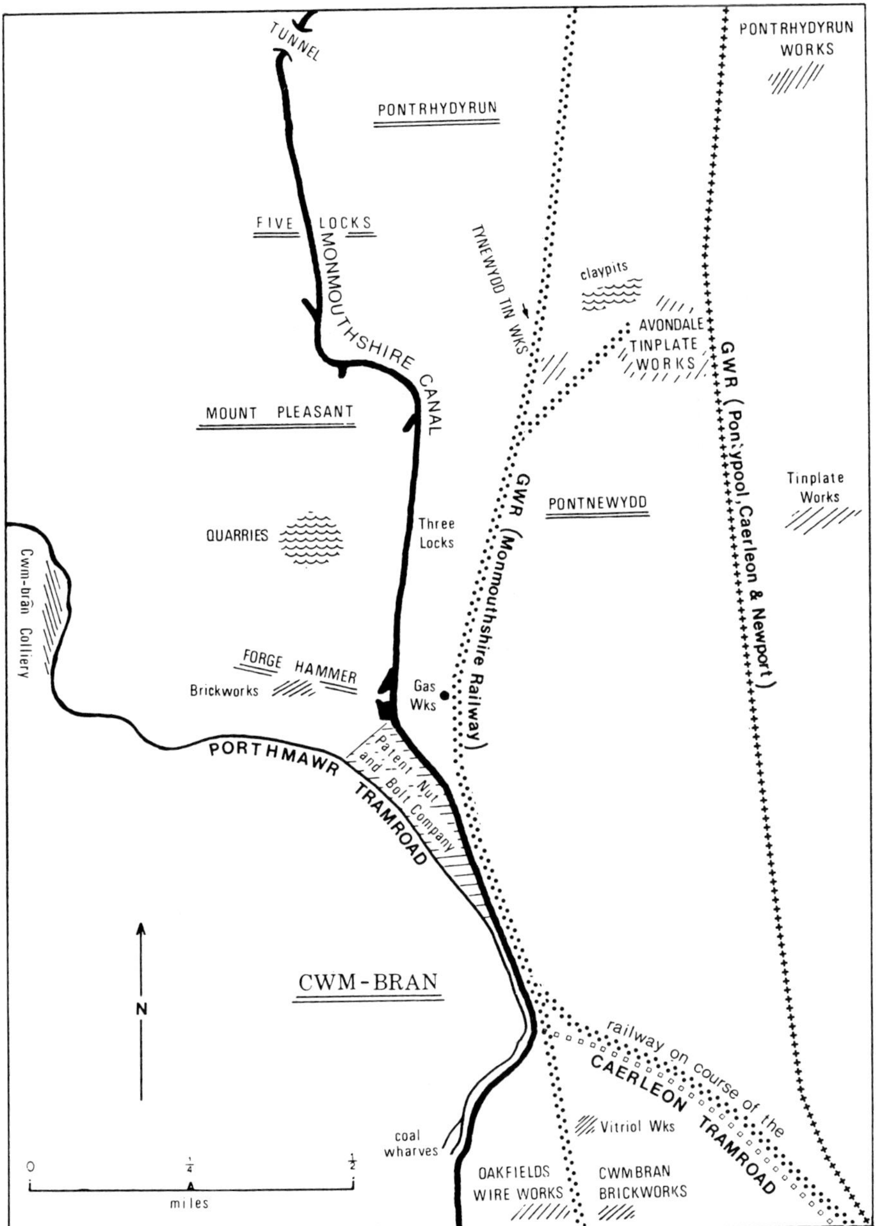

Simplified for clarity

The last market boat on the Monmouthshire & Brecon canal from Moderator Wharf, Newport. The final cargo, carried in George Jobbins boat (trading as Jones & Co.) was loaded on 9th January, 1915: 'the last week for the Market boat to run on the Monmouthshire canal . . . Men joined up and served [being] sent to France'. Note Moderator Sidings signal box behind the fence at extreme right.

Newport Reference Library

ton, iron 20 cwt. of 120 lb. with an allowance of 30 lb. for breakages, thereby proving either the iron was brittle or Welshmen overstrong! However, 'Samuel Homfray and Partners bound themselves to send all Iron, Coal, etc. from their Works to Newport by that line, and the Company agreed to keep the Tramroad in good repair and the Canal well supplied with Water, and in times of frost to keep the navigation open by means of ice-boats'.[2]

Some idea of the enormous output of the ironworks connected with the canal, and as a corollary the traffic thereon, may be gathered from Appendix Five; the tonnage of coal was treble this, while ironstone, tinplate, timber and agricultural produce – including milk, for which express boats ran and fresh vegetables for which a boat left Cefn Wharf at 4 am each day – all added their quota, to give the proprietors a dividend of the maximum 10 per cent allowed by Parliament from 1827 onwards, a position reflected in the price of the £100 shares which rose to a ceiling of £450.

The importance of iron and coal to both the canal and the local population must never be understated. In 1801 a writer says 'The situation of Pont-y-Pool, near a region rich in mineral treasures, in the midst of forges and collieries, and at the head of the canal, render it peculiarly commodious for the establisment of iron manufactories; and perhaps another generation may see a new Birmingham start up in the wilds of Monmouthshire.'[4]

Despite the effects of the high tolls that the Monmouthshire Canal proprietors saw fit to charge (3*d*. per ton per mile, against 1½*d*. on the adjoining and rival Brecon & Abergavenny) prosperity remained theirs until May 1840, when Harford of Sirhowy, no doubt exasperated by transhipment and lockage delays, frightened them sorely with a plan to convert their tramroads to railways. To cap it all in 1842 a plan supported by the Pentwyn & Golynos Ironworks was proposed for a railway from Newport Dock via Pontypool to Nant-y-glo and Blaina. It must be remembered that Homfray of Tredegar had, in December 1829, obtained a 9 ton locomotive from Robert Stephenson & Co. and run it along the Sirhowy Tramroad. After mangling the tramplates somewhat, it petered out in Tredegar Park, the chimney being knocked off; workmen previously sent out to clear the low boughs had at this point retired to the local hostelry; it was a pointer to the future.

However in 1843 the canal company still struggled on with their crude 1 ton trams each carrying a mere 3 tons of coal, without buffers and braked by sprags – a piece of wood or iron jammed in the wheels, a practice which damaged the tramplates (and men) all too often – and loaded and discharged entirely by hand.

On 30th July, 1845, the canal company obtained an Act allowing them to improve tramroads, build a railway from Newport to Pontypool and other works. But no money was to be got, and in 1848 they asked for more time to carry out the work, incidentally changing their name to the Monmouthshire Railway & Canal Company, a change reflected in their canal milestones, the earlier in Newport reading 'M.C.C^{o}', the later, in Brecon – where they were replaced – reading 'MR & CC^{o}'. On 1st July, 1852, the Newport & Pontypool Railway was partially opened, reaching Mill Street, Newport, 9th March, 1853, the canal then terminating at Llanarth Street. Thereafter the story is of less canal traffic and more railway, the MR & CCo. being leased to the GWR

Found in the undergrowth on the Rogerstone flight, this cast iron strapping stump dates back to the days when boats worked through these locks; mute history.

Prior to take-over by the Great Western Railway the Monmouthshire Canal Company had erected combined boundary and mileage markers, as required by their Act of Parliament.

Brecon & Abergavenny half-mile distance post near Brynich aqueduct.

The Great Western Railway were very punctilious over maintaining their boundary line. This is towards the bottom of the Monmouthshire canal, elsewhere stumps bearing dates as early as 1880 are still extant.

NEWPORT AND DISTRICT 1900

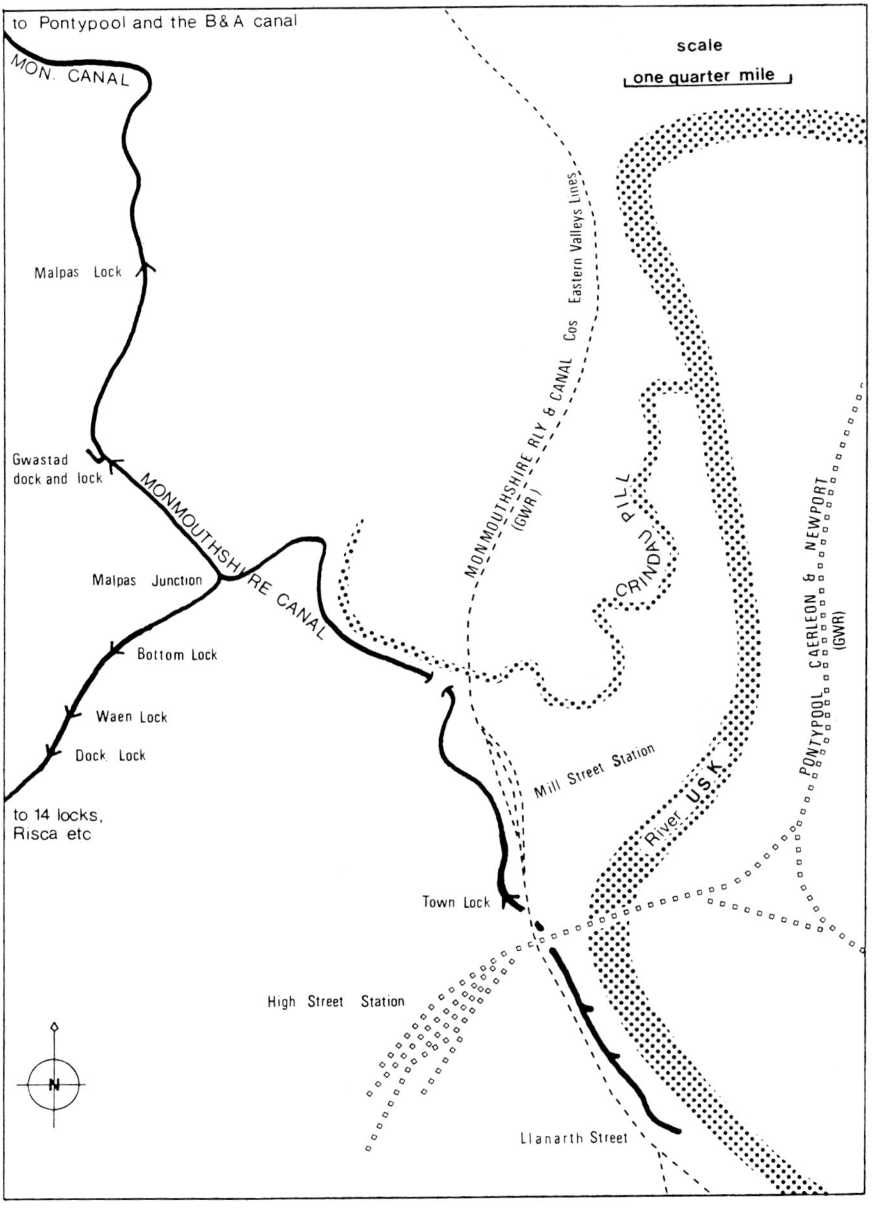

Slightly simplified for clarity

In 1934 the last true traditional South Wales canal boat was built in Tonna yard on the Neath canal. The top photograph shows the framing and bottom planks; the lower view shows the *Ivy May* as she was named just after launching.

(Bottom) The late Will Evans

CANALS OF THE TAWE VALLEY, SWANSEA

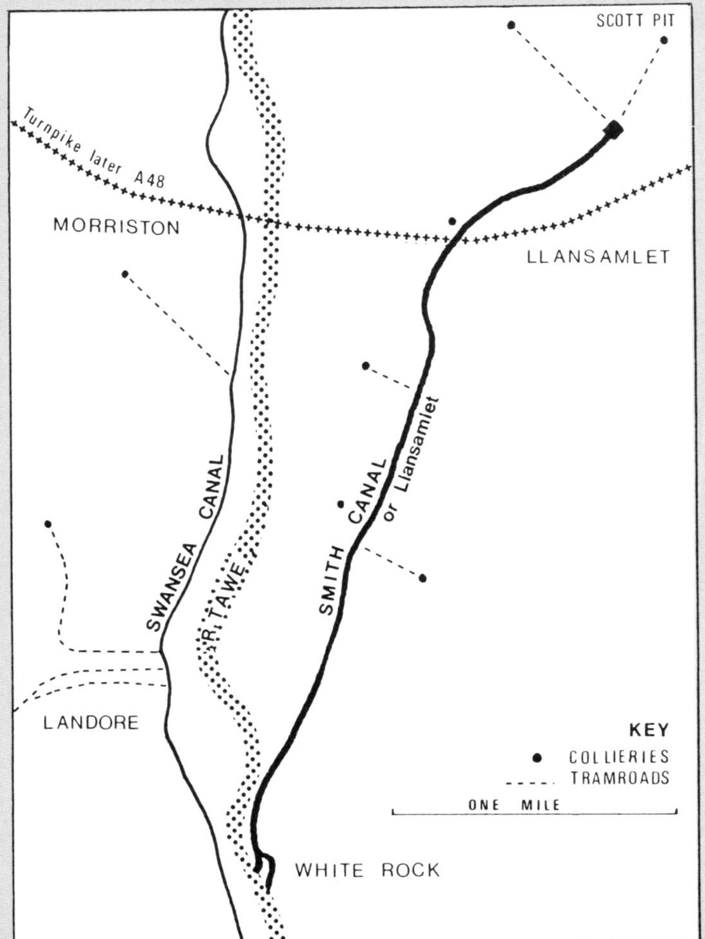

TRANSPORT COMPLEX IN NEATH 1830–1930

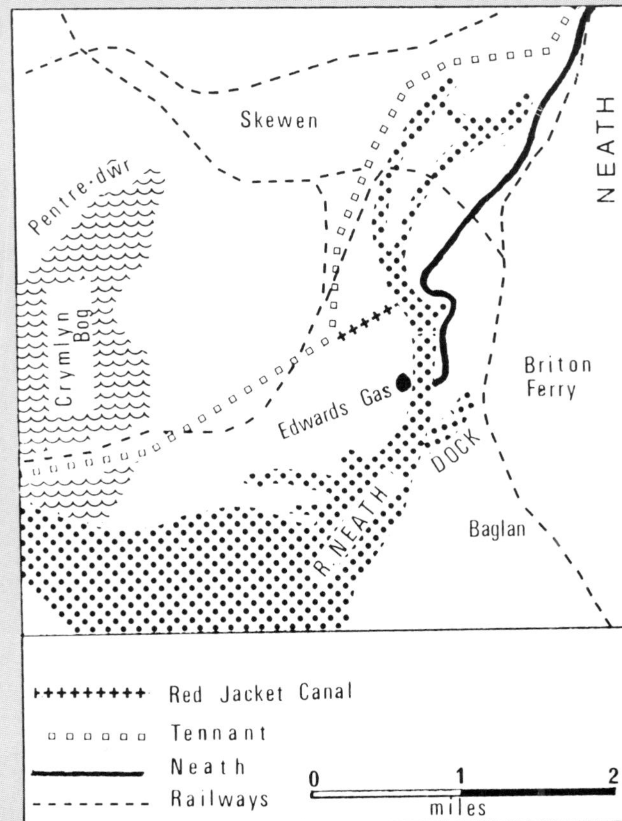

Slightly simplified for clarity

as from 1st August, 1875 and amalgamated with the railway on 1st August, 1880. Curiously, the Sirhowy Tramroad from Tredegar to Nine Mile Point was purchased by the L&NW, in 1876, although a pooling agreement was made with the GWR.

In 1950 Edwards[15] could only say: 'The canal is in a very bad way and most of the locks are unworkable. The weed and general neglect by previous railway owners make this canal scarcely canoeable . . . only section now open is from Cwmbran to Pontymoile.'

THE NEATH CANAL

The first Act for the Neath Canal Navigation was passed on 6th June, 1791 and allowed the building of a canal from Abernant to Neath. A junction was made with the River Neath, upon which there was already considerable trade subject to the usual erratic behaviour of rivers, being flooded in winter and too shallow in summer. Permitted tolls, although a trifle on the high side – from 4*d*. (1½p) per ton/mile for iron to 1½*d*. (½p) per ton/mile for ores and limestone flux – were orthodox enough; but there were exemptions: 'No Tolls are to be paid by the Owners or Occupiers of the Melyn-y-Court Furnace, Ynis-y-Gerwn Tin Mills and Aberdulais Forges, for any Goods conveyed between the last-named Place and the Ynis-y-Gerwn Tin Mills, or between the Weir at the last-named Place and the Upper End of Llyntwrch',[3] and one significant note: 'Pleasure Boats and Boats used for the purposes of Husbandry, are to pass on the Canal and Cuts free of Toll'.[3] As is usual with these canals, the building of tramroads was also permitted.

Originally surveyed by Thomas Dadford Junior, who was then rewarded with the post of 'General Surveyor', by 1792 the work had been taken over by Thomas Sheasby. Arrested due to disagreements with the Glamorganshire Canal Company, he left the scene in 1794 and the canal was officially completed by 1795, the company employing direct labour. In 1798 Rees stated that '. . . this canal was nearly finished, except about two miles at the lower end'.[16] A further Act (26th May, 1798) was obtained to allow an extension to Giant's Grave Pill containing one odd clause:

> And in order to prevent the value of Houses and Lands in the Town of Neath aforesaid, from being lessened by the building of a new Town at or near the termination of the present extension of the said canal, no other house or building other than farm houses and not used for the purpose of any trade or manufactory or as a house of public entertainment shall hereafter be erected or built on the East side of the River Neath on lands of Lord Vernon . . . within half a mile of the termination of the canal, and if so built shall be deemed to be a common nuisance and be prosecuted as such.[1]

Although the survey was undertaken by Dadford by 1799 the work was carried out using a Welsh contractor/engineer Edward Price. In the meantime Sheasby had reappeared and undertaken some work on the main line, enabling the whole to be ready by the opening date (1799) at an overall cost of £40,000 for 10½ miles and 19 locks.

A Neath Canal bridge in 1914. The gas lights and telegraph pole add a piquant touch; the bridge looks to have been recently re-pointed. *British Rail*

In the Vale of Neath, *c*.1908. This length of the Tennant Canal near the Mile End Road, Swansea, was infilled in the late 1950s, although the cottages are still extant. *City of Swansea Archives*

Two views of Glyn Neath (or Glynneath). Always plagued by silting (as long ago as 1811 following a Court case the company stated 'the Canal would shortly have been choaked up [and] the Trade annihalated'), this part of the waterway was disused by 1921. The canal basin appears to be busy enough. The proximity of the terraces to the railway marshalling yard is very apparent in the top view. *Pontypridd Public Library*

Perhaps one of the most famous bridges on South Wales canals is Pontgam (the crooked bridge) at the junction of the Neath and Tennant canals at Aberdulais. The photograph shows the canal in 1914 with the Neath bearing to the left and the Tennant swinging to the right. *British Rail*

Aberdulais Works like the Neath canal required considerable labour; typical canalside housing of terraces, *c*.1920.

On 8th May, 1948 a weedcutter is seen at work near Aberdulais. Although by this time navigation had virtually ceased, the importance of both the Neath and Tennant canals as water supply channels justified considerable maintenance work. Left to right, W.P. Jones, H. Jones, Ted Bolt.

A few years later and showing more detail of the paddle wheel drive of the boat.

This is believed to be Will Evans, for many years the foreman of the Neath canal; this photograph taken in the 1930s at a time when the canal was dewatered serves to illustrate the lot of canal workers past and present.

Paying a dividend on average of 15 per cent up to 1850, the canal was easily able to resist the temptation of railway ownership although even in 1845 we find: 'The Neath canal, which passes through a district abounding with mineral wealth, terminates at this place, after a course of about fourteen miles; the wharfs are at a place called Giant's Grave, where upwards of of sixty thousand tons of coal and culm have been shipped annually, but there are no works at present in operation, with the exception of two or three coal pits, in which a few hands only are employed.'[4] After this date a decline set in, as traffics, mainly from the collieries, moved away and the final dividend was paid in 1896. After falling to a mere 11,000 tons per annum in 1907, commercial traffic ceased in 1916 when a market boat ran through, the last boat working back empty from Glyn-Neath in 1921, although a new maintenance-cum-carrying boat was launched in August 1934 at Lockhouse, having been built from scratch by the carpenter Ben Jones and his gang.

THE SMITH CANAL

The Smith (otherwise known as Llansamlet) Canal came into being when on 22nd February, 1793 the Duke of Beaufort granted permission to John Smith (Chauncey Townshend's son-in-law) to construct a canal from Llansamlet to White Rock. Coal had been worked in the Llansamlet area as early as 1707–8 and taken on the backs of mules to the River Tawe at the White Rock. In November of 1750 Chauncey Townshend leased the coal lands on the east side of the river from Mansel and a month later the Duke gave Townshend leave to construct wagon ways from LLansamlet to White Rock.

By 1784 the canal had been made in part, running parallel with, and in parts superseding, the wagon way. The canal being completed in 1792 served two purposes, the exportation of coal via the River Tawe and the short haul traffic of coal for consumption in the Copper Works on the east side of White Rock, Middle Bank and Upper Bank. Smelting and refining processes were at this time highly secret and if you were a Marfa man you did not speak to a Hafod man lest you let slip one of the 'mysteries of copper smelting' which both factories hid well behind high blank walls.

THE SWANSEA CANAL

Falling no less than 373 feet via 36 locks in a total mileage (excluding branches) of 16¼, the Swansea canal received its assenting Act on 23rd May, 1794. Commencing at Hen-neuwadd it finds its outlet to the sea at a junction with the North Dock. Underlying these simple statistics is the story of a successful canal, albeit one that has, save a very few traces, totally disappeared.

On 18th March, 1793, it was reported that several of the burgesses and aldermen on the Swansea Corporation, at a meeting, declared: '. . . it appears to us that . . . a canal will be of very great utility to all places within reach of the same, that it will be the means of very greatly improving the Lands and Manufactories of all its neighbourhood and greatly enriching all the Country'.[17]

The Swansea canal at Clydach showing disused works, chapel, warehouses, dog and, surprisingly, English graffitti in 1973.

The Carmarthenshire Tramroad (or Railroad as early as 1803) was short-lived, the 13 miles being completed in 1804, and disused, with the holding company in liquidation, by 1844. It ran from the docks on the Burry River at Llanelly to collieries and limestone works in the area of Castell-y-garreg near Llanybie. This early engraving is dated *c*.1820. *Waterways Research Centre*

COAL, IRON & WATER IN THE INNER VALLEYS, with works &c. indicated, AS SURVEYED IN AD1842

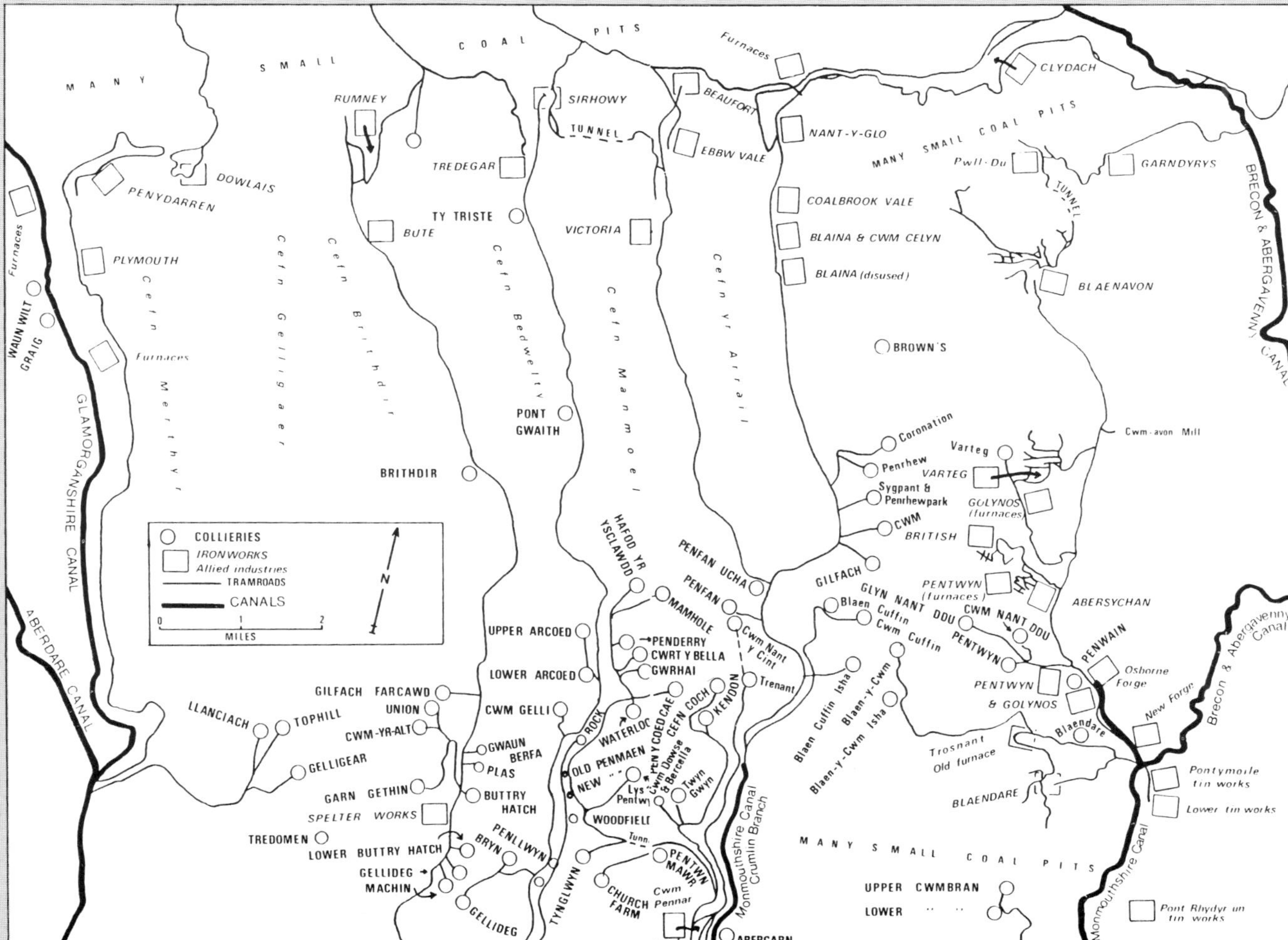

The first public meeting took place on 5th April, 1793 when Thomas Sheasby was appointed surveyor. During the period between this and the obtaining of the enabling Act, feverish discussion took place as to where in Swansea the canal was to terminate, which resulted in the Duke of Beaufort building his own waterway (the Trewyddfa) from Nant Rhydyfiliast to Nant-y-felin, albeit as a part of the Swansea main line, and collecting his own tolls. But if we are to believe a letter from Thomas Morgan, received 31st March, 1794, the Duke was not intent on feathering his own nest. '. . . You may depend upon it the Duke and his friends are as eager for a canal to the benefit of Swansea or were so, as any . . . The artful designs of wicked men labouring to serve themselves at the ruin of others can prevent it . . . There are too many who act on no other principle than their own advantage, and who openly declare they care not for the canal or the Publick if they can be served . . .'.[17]

Although Sheasby was in nominal control, on 15th August, 1794 a local man, Charles Roberts, gave his report on the plans for the waterway, which envisaged a narrow canal taking boats ('barges') 70 ft × 7 ft, carrying 20 tons a-piece; the canal to be 30 feet wide except 'where the ground was incommodious', where it would be narrowed, and five feet deep to give a good reservoir of water; but at the summit the width should extend to 40 or 50 yards wide, as water could only be drawn from the main feeder, the River Tawe, on Sundays, and Mr Roberts feared that 'supposing the canal to be full on Sunday evening and no more water to pass into it before the following Sunday, I am apprehensive that so much water would be drawn out on the first two days as to render the other four less useful than they ought to be; this circumstance can only take place in dry seasons, and upon the upper part of the canal'.[17] A further complication was that the Tawe, whose valley the canal would share, 'frequently outreaches its common bounds and trespasses very considerably on the adjoining lands'[17] and it was necessary to carry the canal five to six feet above the level of the wash meadows.

Notwithstanding these difficulties the canal was opened to Godre'r-Graig in 1796 and throughout in October 1798 at a cost of about £55,000, rather less than the authorised capital.

Success was quickly achieved and when in 1802 an early gongoozler* appeared, he reported that,

> By the side of this canal the walk is very pleasant, affording many objects worth observation. There are no less than 36 locks in the space of sixteen miles, from an elevation of 372 feet, and several aqueducts. Adjoining are large smelting copper-works, the iron-forge, brass and tin-works, a fine copper rolling mill, iron furnaces and foundry, and a most stupendous steam-engine at Llandwr, which cost the proprietors upwards of £5,000 to complete; this machine throws up from a vast depth one hundred gallons of water each stroke, which is repeated twelve times in a minute, making 78,000 gallons of water in an hour; it was made by Messrs Boulton and Watts',[18]

and in 1804, 54,235 tons of coal and culver were moved with tolls paid of £3,590. In 1805 an unfortunate breach of the canal bank obstructed the harbour, as 1600 cubic yards of gravel and debris was carried into the river

* One who idly watches the canal and its doings.

just above Harper and Sheasby's wharf. The underlying cause was a weak retaining wall at the wharf, and while discussions were carried on as to who should remove the shoal, boats trapped above had to face delays of 6–8 weeks.

Notwithstanding this and other minor contretemps – including complaints from traders of the poor condition and low water levels in the canal – dividends of between 10 per cent and 16 per cent were paid from 1819 to 1848, although as usual when dividends were in any danger of falling wages were cut; typically a lock-keeper earning 18s. (90p) a week in 1807 received only 15s. (75p) in 1816. Tolls were also cut, but on the return of prosperity they again rose.

In 1845 Joseph Bailey – brother to Crawshay Bailey of Aberaman (Aberdare) Ironworks – was Chairman of the Welsh Midland Railway, whose intended route was from near Worcester via Hereford and Brecon to Merthyr with branches through the valleys and hence to the ports. The Swansea shareholders promptly agreed to lease their canal to this concern for 99 years at £8 p.a. per share. The whole scheme died in 1846 when the Welsh Midland failed to get its Act, but the canal entered a period of prosperity; mines, iron-foundries and copperworks were all at full blast and as a result dividends of 18 per cent were paid from 1856 to 1860. This was an 'Indian Summer', for in 1864 the Swansea Vale Railway was opened and both dividends and tolls fell, the former to a mere 8 per cent in 1865, while the older ironworks of Ynysgedwyn and Ystalyfera, unable to compete with the products of modern steelworks, first cut their trade and later ceased completely, the former closing in 1876 and the latter ten years later.

In 1871 an agreement was sought and made with the Great Western Railway, the whole waterway being sold for £107,666 in shares – £202 per £100 share – with the Duke of Beaufort picking up £40,000 for his Trewyddfa Canal. The sale was dated 31st January, 1873 and to all intents and purposes gave the shareholders a permanent 10 per cent dividend.

Built by Isca Foundry Newport in 1894 this overbridge carried the Rhondda & Swansea Bay railway over the Tennant canal. Photographed in 1912.

Canals Officer, Neath Borough Council

Until 1894, however, the canal continued to be 'in the black', the GWR, having no nearby railway line, was glad to work it hard and hence to poach traffic from the Midland Railway, but after 1912, with a trading loss in that year of £238, the deterioration set in and the final end of traffic in 1931 caused no surprise.

THE TENNANT CANAL

The Tennant Canal was a logical extension to the Glan-y-Wern and Red Jacket Canals and was propounded by George Tennant in 1821, with William Kirkhouse as his Engineer. No Act was necessary as the local landowners were all presumed to be in favour of the scheme. Four and seven-eighths miles long, it runs from the Neath Çanal at Aberdulais to make a junction with the Red Jacket Canal.

Kirkhouse, described as 'a bluff independent personality, strong in the knowledge of his abilities and confident in his powers'[19] needed to be all of this, for near the Neath Abbey Ironworks a steep cutting 500 yds long and 30 ft deep was required in a quicksand 'of the finest grain and most subtle nature'. No matter how the navigators dug, and despite a 45° batter on the sides, the sand would not stop shifting until an inverted arch of masonry was introduced as a retainer. It was fortunate, too, that Tennant was a rich man, as in order to make the junction with the Neath Canal a 'stupendous and magnificent' ten-arched aqueduct was necessary which, although commenced in May 1823, was not completed until a year later.

On Whit Monday 1912 every Sunday School in the district must have taken part in this outing on the Tennant Canal, seen near the Swansea bridge

City of Swansea Archives

Following a breach near Pontypridd essential supplies had to be trans-shipped at Foundry bridge, Upper Boat. The cabin boat is from the Glamorganshire canal company's own carrying fleet; October 1908. *Canals Officer, Neath Borough Council*

A Sunday School outing on the Tennant canal around the turn of the century. Everyone appears to be having a super time. *Canals Officer, Neath Borough Council*

The opening ceremony was carried out on 13th May, 1824, with some aplomb:

> Then barges fully laden with coal were safely conveyed from the colliery of Messrs Protheroe in the Vale of Neath, and delivered at the shipping place in our [Swansea] harbour. A fine pilot boat from Briton Ferry, named the *Countess of Jersey*, led the van, bearing several musical performers. A handsome pleasure boat followed, having on board Mr. E. Waring, private agent, and Mr. Kirkhouse, the able engineer of the concern, accompanied by several gentlemen of Neath and Swansea. Then came a barge appropriate to the carpenters and masons who had provided themselves with a sort of floating battery for the occasion, which they discharged at intervals by way of salute with great regularity and good effect. All the boats were handsomely decorated with flags.
>
> At Neath Bridge a large river barge was substituted for the pleasure boat, and amidst the acclamations of a dense crowd of spectators, a vast number of whom accompanied the boats, some on the towing path and others on board, the line of vessels proceeded, being joined near the Abbey by two barges belonging to Mr. Parsons and laden with coal. The whole arrived at Salt House Dock at 2 o'clock and was welcomed by the firing of guns, ringing of bells, and the display of colours from numerous vessels in the Port . . . .[17]

Tennant's workmen were also royally treated and the toasts to Tennant included: 'May his canal return him a large interest for all the capital expended on it'.

That the canal was popular with the natives of the area is undoubted as witness the 'address' given on the presentation of a gold snuff box on 12th October, 1824 to George Tennant by D. Morgan, Portreave of Neath:

> Sir, in the names and on the behalf of numerous Tradesmen in the Town of Neath and its vicinity, we wait upon you to request your acceptance of this Snuff Box. We offer it to you both as a tribute of our respect and as a testimonial of our gratitude for the benefits which you have already conferred upon this neighbourhood . . . We beg leave further to offer you our hearty congratulations upon your recent completion of so great an undertaking, and to express our sincere wishes that you may be blessed with health and length of days to enjoy the fruits of an Enterprise which cannot but prove eminently and permanently useful.[1]

Not content however with having a canal known by his name, although he admitted that he now and then indulged in 'a secret vanity that his name might hereafter be remembered in Glamorganshire as one who had caused a scratch to be made upon its future map . . .', he also had built a dock at the seaward end which encouraging traffic gave the canal an annual tonnage of 85,000 by the time of his death in 1832.

The boats used, similar to those of the Neath Canal carried about 25 tons and although only requiring 30 inches of water to swim in, had a 5 ft depth to Red Jacket to guard against lack of water supplied in summer. The maximum tonnage carried on the waterway was roughly 225,000 in 1866 and although this dropped gradually over the years as the railways were opened, nevertheless even until 1895 the canal continued to be profitable, maintaining a steady 100,000 tons p.a. By 1904 the maximum draught was shown to be 2 ft 9 in. due to a lack of dredging which restricted loads somewhat and, as was the case in the Neath Canal, boat traffics ceased in the 1930s.

Four views of Aberdulais aqueduct carrying the Tennant Canal over the river Neath. The top illustration on this page shows the structure taken from the river bed, and the lower, Lock No. 1 on the Tennant canal and the toll house. Both photographed in 1973.

The aqueduct viewed from behind the arches of the parallel railway viaduct and the virtually de-watered channel were both photographed in 1989. The first stones were laid in May 1823, the Engineer, William Kirkhouse, seeing its 340 ft (103.6 m) length completed by April 1824. The fact that it is not unknown for the river to rise over the top of the aqueduct is a salutory indication of its seasonal force.

Chapter Two

Minor Canals

Gone and (almost) forgotten is the best phrase to use in describing some of the odder, often unfinished, canals of the valleys. One of (if not the) earliest canals was a short cut that ran from Aberdulais to the River Nedd, thus providing a transportation method for iron and tin from the Ynys-y-gerwn Mills to their 'old works' at Aberdulais. Possibly opened in 1750 it was certainly extant in 1768 when a lease from George Vernon included '. . . Navigable Cut thro' Aberdylais Vach Farm from the Lady Day 1768 for 21 years & Renuable as often as required in the Life time of the said Mr. Vernon. Yearly Rent. £5.'[1] It became disused at the end of the eighteenth century.

The *Dafen Pill Canal* was, similarly, merely a short cut (¾m.), made to Chauncey Townshend's collieries at Treinon. Opened *c.* 1770 it was still in existence, probably unusable, in 1830. Of equal life was the *Yspitty Canal*, another brainchild of Chauncey Townshend. About a mile long it ran from Yspitty to a colliery near Bynea with a branch to near Pencrug.

The '*Englishman's Ditch*' has only one claim to fame, and indeed little proof of its existence, save that in a copy of Oldsworth's *Swansea Guide* held in the library of the Royal Institute of South Wales, additional notes by Mr Rees Jones of Loughor tell us that across the Crymlyn bog there was '. . . an old canal . . . It is not exactly known at what time this was cut, but on clearing the ditch the present adventurers . . . found the skeleton of a boat which had, in all probability, been anciently used for the conveyance of coals to the burrows from the collieries in the neighbourhood, from which circumstance it appears that the exportation of coals was carried on here at an early period . . .'[1]

The '*Cnel Bach*' or Little Canal was a branch of the Neath Canal from near Aberclwyn to the river, at which there was a tipping place to which Cwm-gwrach coal and iron was brought by a tramway. Built about 1814, the canal remained in use until 1854. Curiously when the bridge over the river became too dangerous to use – about 1880 – the children from Rheola instead of going to school at Cwm-gwrach had perforce to go to Resolven. The most famous moment in the life of this canal came in 1824 when ten boats laden with Cwm-gwrach coal passed through *en route* to the grand opening of the Tennant.

The *Penclawdd* was a respectable waterway (Act 21st May, 1811) built within the estimate – including tramroads – of £9,934. Three and five-eighths miles in length, it commenced at the 'New Quay' alongside the River Loughnor. From there it ran east to near Ty-dan-y-lan and then north for ¼-mile. Swinging east again it took a rather irregular course to Pont-y-Cob and then more or less straight to the Elba Steel Works, Gowerton. Once more bearing east it ran along the south side of the River Llan to a ¼-mile west of Ystrad Isaf Farm where it met its tramroad to the Waunarlwydd Colliery. With only two locks, it was opened around 1812 but in 1845: 'At the village of Pen-clawdd . . . were formally extensive copperworks belonging to the Cheadle Copper Company; but they are now neglected, that company possessing numerous others in more convenient situations, and are going to ruin',[4] as had the colliery in 1814, the canal closing shortly thereafter. Incidentally, the permitted rates including 4*d*. (1½p) per ton per mile for iron and 2½*d*. (1p) for pig iron were comparable with those on the Neath and Swansea Canals, both of which latter had far heavier engineering works.

The *Cyfarthfa Tub-boat Canal* was an oddity. Built at the behest of Anthony Bacon about 1778, it took boxes (each carrying about 2 tons) in trains of six, hauled, or more probably shafted, by one man from the collieries at Cyfarthfa to the works. 'The whole is upon one level, and it does not connect with any other canal or navigation; it is now the property of Mr. Crawshay, the great iron-master. It is situate near to the northern ends of the Glamorganshire canal . . .'.[16] Closed around 1835, some of its underground connections – leats to the pit-workings – may still be in existence.

The *Doctor's Canal* (variously Dr Thomas's or Dr Griffiths), built in 1809, was only one mile in length from Denia to Treforest. Four years later in a physical connection with the Glamorganshire Canal was permitted at Denia at the tail of lock 34, but of more importance was the connecting tramroad to the collieries at Gyfeillon (The 'Hafod' q.v.). Described in 1904 as 'having a small trade . . . from a colliery [Pwll-Gwaun] communication with Treforest Wharf by a tramroad',[10] it failed to survive World War I being described as 'derelict' in 1918. The bed may still be extant.

The *Earl of Ashburnham's* was a purely private venture of roughly 1½ miles from his collieries at Ffrwd, Coed (via a branch) and (by tramroad) at Pembrey, to an outlet on the south bank of the River Gwendraeth. Opened 1796, it was still extant although little used in 1847.

Of *Kidwelly*, it was said in 1845:

> Some docks, and a short canal, were constructed here about the year 1766 by Mr Kymer, a private individual; the former are situated about half a mile from the town, and the latter was intended to convey coal from the mouth of the neighbouring pits to the vessels in the harbour. Some years since, this navigation was transferred to a company known as the Kidwelly Canal Company . . . [and] . . . it now extends for fifteen miles. Here were formally both iron and tin works, the former of which have been entirely abandoned, and the latter are now conducted only on a very limited scale. The only exports are coal . . . and corn, and cheese . . . .[4]

The first Act was obtained on 19th February, 1766 by Thomas Kymer to cut his canal through his own land, from both his collieries at Pwll Llygod and their adjacent lime kilns to the Little Gwendraeth River. Opened in 1769, 'the utility of the work, however, was proved to be so great, that it was thought desireable to extend its benefits to the neighbourhood, and, in consequence, another Act was obtained in 1812 . . .'.[3] This allowed for the building of a waterway connected with, and incorporating Kymers Canal, from Kidwelly to Llanelly, a further canal or tramroad up the Gwendraeth valley to Cwm-y-Glo and other branch tramways or canals. Permitted tolls varied from 4*d*. (1½p) per ton per mile for general goods, 2½*d*. (1p) for pig iron to 1¾*d*. for 'rottenstone', coke and bricks. Harbour dues were to be 1*d*. per ton burthen but *inter alia* 'His Majesty's Vessels of War' were exempt.

Conversely. 'For every Horse, Mare, Gelding, Mule or Ass, Cow or other Neat Cattle, travelling on the Railways or Tramroads, and not drawing any Waggon, nor passing from one Farm to another on the Line, 1*d*. each.' James Pinkerton was the Engineer and by 1824 the line making connection with the Earl of Ashburnham's Canal near Ffwrd and also from Spudders Bridge a channel was dug as far as Moat Farm, with a tramroad connection to Trimsaran colliery.

In 1832 following the completion of Pembrey Harbour (now Burry Port) the Kidwelly and Llanelly company asked James Green, possibly the most ingenious and greatest canal engineer, to report on their position. *Inter alia* he suggested that a line of canal be made from Pen-y-bedd to the new harbour, that a connecting tramroad be built from the canal at New Harbour, to the Carmarthenshire Canal and that – a magnificent feat of engineering – the canal through the Gwendraeth valley be extended to Cwm-mawr, utilising three inclined planes, with a total rise of just under 200 feet.

Of his three proposals, the Harbour connection was in use in 1835 and the tramroad by 1837. The valley canal was also in use at this date, although according to Charles Hadfield[24] the topmost incline at Hirwaun-Isaf was never completed, the colliery owner at Pontyberem utilising a tramroad along a towpath and incline instead. As with the Cyfarthfa Canal, tub-boats albeit of a 5 ton capacity, were utilised on this section. They may well have been wheeled and run straight up the tramroad tracks. So at a cost of around £75,000 the company had about 20 miles of tramroad and including the lease-held Kymers Canal, 12⅜ miles of canal. The first dividend (of 1½ per cent) was paid as late as 1859 when, of course, the waterway was clearly outdated. This was also apparent to the shareholders, and after beating off some opposition, they converted themselves in 1865 to a railway company, subsequently (1869–91) building lines over most of their waterways and tramroads.

In the meantime,

> To facilitate the conveyance of the mineral produce of the district, a capacious harbour was constructed in 1819 by the *Pembrey Harbour* Company . . . but this harbour has fallen into disuse, being private property, and has been superseded by another construction to the north-east of it . . . a canal has also been formed, connecting this port with the Kidwelly and Llanelly canal, and pursuing hence a northerly course; and Pembrey, in consequence of these and other advantages, promises to become a place of great trade[4]

a hope not to be realised. Described as a 'useful artifice' the waterway was eventually extinguished by the track bed of the South Wales Railway.

It is perhaps apposite to finish with a magnificent 'might have been', the *Grand Towy* Canal.

> A canal of great undertaking in this part of Caermarthenshire has been, off and on, in projection these 30 years . . . intended to commence at the port of Loughhor, and cross the country through or near the collieries, mines, and limeworks of Llandebie, to the Vale of Towy near Llandeilo; and up that vale to the neighbourhood of Lord Cawdor's extensive and rich lead mines. This canal promises great advantages to the adventurers, and will be productive of infinite goods to the

> neighbourhood at large, by conveying fuel and manure to those parts that now stand in need of them; and affording a ready outlet to the mines, coals, and other productions of the country . . . Unfortunately water, too cold even for a canal, was thrown upon the scheme, and it was for the time abandoned . . . When Peace, now upon its march, returns from abroad, may all be peace an unanimity at home; and in that favourable crises, may the Grand Towy Canal be taken earnestly in hand, and completed; to the certain emolument of the subscribers, and the no less certain and incalculable improvement of the whole tract of the Vale of Towy above Llandeilo, and its vicinities; even the hundred of Buallt in Brecknockshire, and the eastern parts of Cardiganshire, would experience the beneficial effects of reduction in the prices of fuel and manure; and of a cheaper conveyance of the produce of the country to a sea-port . . . The whole length of this line of canal, with its branches, might be extended to from 50 to 60 miles.[6]

Now that would really have been worth exploring!

Chapter Three

Canal Tramroads

Tramroads, simply, were precursors of our modern railways. The initial form of longitudinal wooden rails resting on transverse sleepers was recorded as being in use in Wales as long ago as 1695. Of beechwood, the rails were quickly worn, and in 1738 iron plates were nailed to them to reduce this wear. From that composite form it was a short step to the use of iron throughout. Naturally each engineer had his own ideas on construction; in 1825 two styles of rail were described by George Overton, Engineer *inter alia* of the Aberdare Canal and the Penydarren Tramroad.

> On the edge-railway the rims of the carriage wheels are made broad, with a flanch on the side of the wheels to prevent the carriage from leaving the rails, which renders the carriage or wagon considerably heavier than the tram . . . [alternatively] the wheels of the tram are narrow in the rim, coming nearly to an edge; and there is a flanch on the inner side of the plate upon which they move, to prevent the carriages or trams from running off the road.[20]

The building of a tramway in 1798 was seen by a visitor who wrote a clear description:

> In the vicinity of Blaenavon we observed the process of making a railroad, so called because it is formed by a kind of frame with iron rails, or bars, laid lengthways, and fastened or cramped by means of cross-bars. The ground being excavated, about 6 ft in breadth, and 2 ft in depth, is strewed over with broken pieces of stone and the frame laid down; it is composed of rails, sleepers, or cross-bars, and under-sleepers. The rail is a bar of cast iron, 4 ft in length, 2 in. thick, and 1½ in. broad, its extremities are respectively concave and convex, or, in other words, are morticed and tenoned into each other, and fastened at the ends by two wooden pegs to a cross-bar called the sleeper. The sleeper was originally of iron, but experience having shown that iron was liable to snap or bend, it is now made of wood, which is considerably cheaper, and requires less repair. Under each extremity of the sleeper is a square piece of wood, called the under-sleeper, to which it is attached by a peg. The frame being thus laid down and filled with stones, gravel, and earth, the iron rails form a ridge over the surface, over which the wheels of the car glide by means of iron grooved rims, 3½ in. broad. This is the general structure of the road when carried in a straight line; at the junction of two roads, and to facilitate the passage of two cars in opposite directions, moveable rails, called turn rails, are occasionally used, which are fastened wuth screws instead of pegs, and may be pushed sideways. The level of the ground is taken with great exactness and the declivity is in general so gentle as to be almost impercept-ible . . . The expense of forming these roads is very considerable, varying according to the nature of the ground, and the difficulty or facility of procuring proper materials; it is seldom less than a thousand pounds per mile, and sometimes exceeds that sum.[2]

Elsewhere and more commonly, stone blocks were used as sleepers. Benjamin Outram, in 1799, recommended these to the proprietors of the Brecon & Abergavenny Canal, describing the method of construction as:

> The ground is first formed in the best manner the nature of it will admit, and perfectly drained; then covered with a bed of small stone or good gravel, six inches

Blaenavon 'Big Pit' is now a tourist attraction with experienced miners taking visitors 300 ft to the pit bottom. On return to the surface you will find much remains of the tramroads of the area; and some idea of the industry that once was, can be gained.

Reproduced from the 1902, 6" Ordnance Survey map

> in thickness, and four yards in breadth, for a single road, and six yards for a double one; on this bed, stamped firm, are placed blocks of stone for sleepers for the rails; each block being upwards of one cwt.; in the centre of each block a circular hole is drilled six inches deep, and in it is put an octagonal plug of oak five inches long, which receives an iron spike, that fastens down the ends of the two rails that rest upon the block, the spaces between the blocks are filled with small stones, which are rammed close about the blocks, and covered with gravel, but not so high as the soles of the rails outside, nor so high as the top of the flanches on the inside between the rails . . .[6]

In Wales extensions to canals were rarely considered, tramroads having the advantage in hilly country; but due to the hills inclined planes were often necessary. Normally these were of a counterbalance type, the laden downward trams hauling up the empty ones. There were objections to this method of working: delay whilst assembling full and empty trams, and accidents: 'A continual expense in wear and tear of ropes or chains, attendance of men, &c., and we may therefore establish it as a general rule, that the inclined plane should never be adopted, except in cases of absolute necessity.'[20] Where such lines existed – Hill's tramroad from Blaenavon via Garndyrys to Llanfoist for example – extra stones were laid to carry the ropes or chains; deeply scored, they are easily distinguishable from the normal blocks.

In many cases tramroads were built, served their purpose and died, often due to the closure of outmoded ironworks or of out-worked quarries. Some indeed were superseded by railways, but others, often for reasons of convenience, worked side-by-side with the usurper.

Brief details of most tramroads built in the period 1760–1840 are listed below in simplfied form, including where possible the following information: name of the tramroad, terminal points, mileage, Engineer (NK = Engineer Not Known) and operational dates. Closing dates are either the date of total disuse or of conversion to an orthodox railway. Lines of less than 2 miles in length, unless of special interest, are excluded.

Aberdare Canal:

Abernant	From Hirwaun Ponds to Abernant Ironworks (4 m.); E. Hopkin, 1805–1900.
Llwydcoed	From Canal Head, Aberdare to Abernant Trd, Llwydcoed (2 m.); G. Overton, 1811–1900. 4 ft 2 in. gauge.
Penderyn	From Bryngawn Patch to Penderyn Quarries (2 m.); T. Dadford Jnr, 1794–1844.

Brecon & Abergavenny Canal:

Bailey's	From Govilon to Brymawr and Nant-y-Glo (5½ m.); NK, 1822–1859.
Beaufort	From Gilwern to Beaufort (5¾ m.); P. Williams, 1796–1860 & 1908 (numerous branches).
Brynoer	From Tal-y-Bont on Usk to Trevil & Rhymney Ironworks (12 m.); G. Overton, 1815–1865.

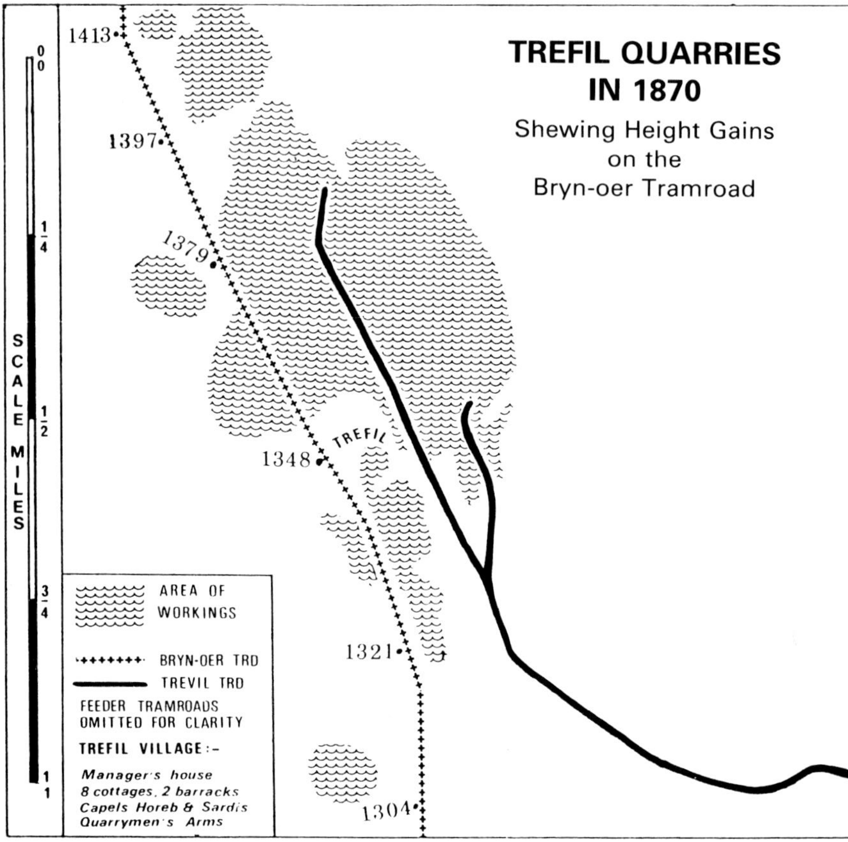

Slightly simplified for clarity

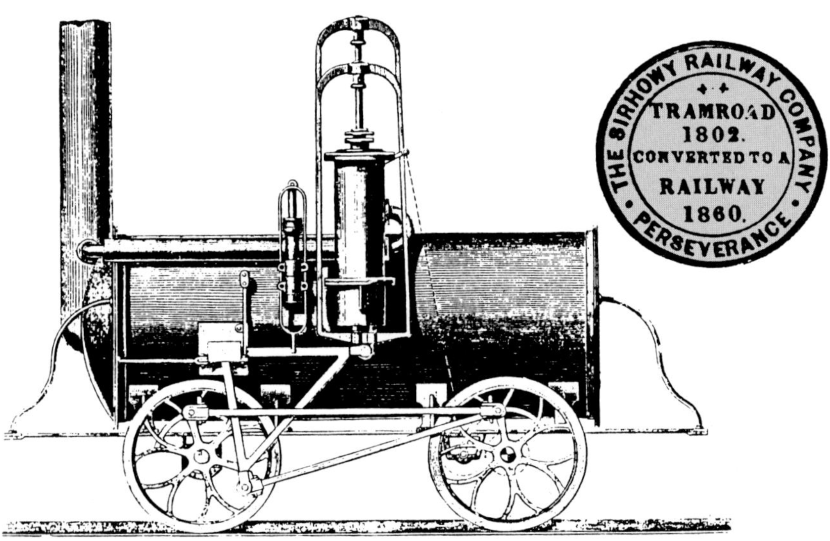

The *Speedwell* built in 1860 by the Neath Abbey Iron Works for Thomas Prothero of Newport for service on the Sirhowy line. *Drawing courtesy Railway Magazine*

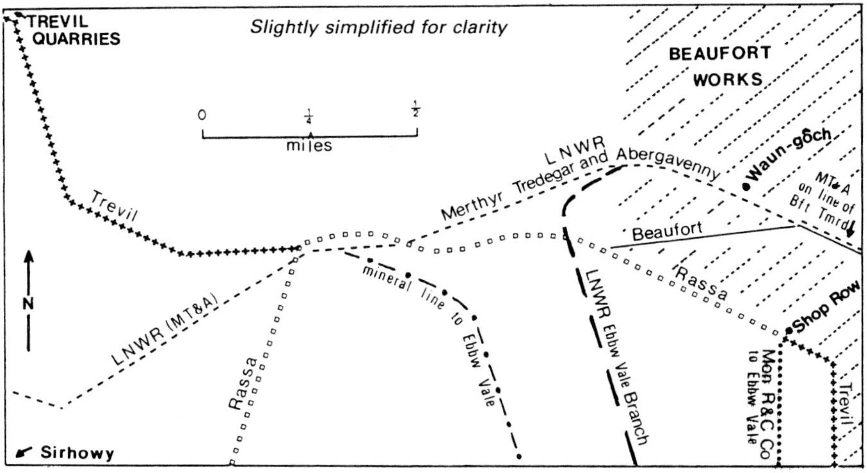

RAILS (and plates) TO BEAUFORT

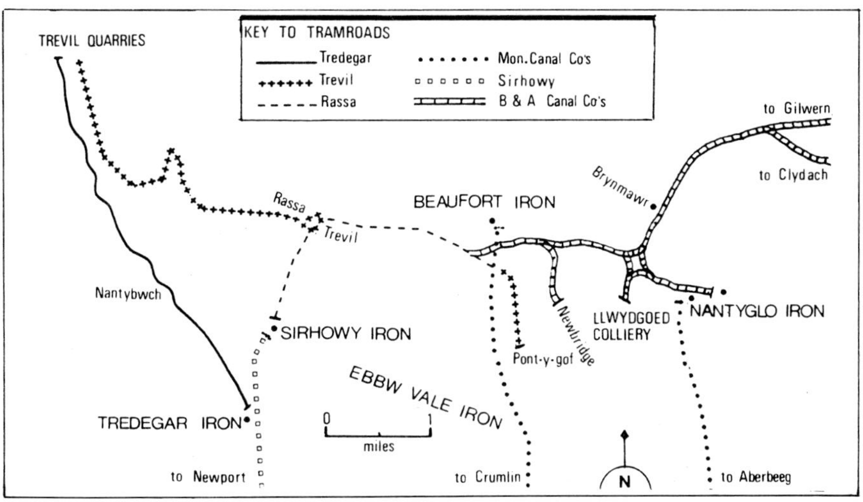

Graig-yr-Hafod	From Llanover to Graig-yr-Hafod (3½ m.); NK, *c.*1799–*c.*1850.
Grosmont	Continuation of Llanfihangel Rly at Llanfi Crucorney to Llangua Bridge (7 m.); W. Crossley, 1818–*c.*1852. Act 20th May, 1812. Est. £12,000. 'When it is stated, that by this railway a difference in the level of 166 to 168 feet is made in the distance above specified, it is hardly necessary to add, that were it even for nothing but the saving of time and labour in the conveyance of goods, the work could not fail to be of very great utility.'[3]
Hay	From Watton Wharf, Brecon to Junc. w. Kington Rly, Eardisley (24 m.); J. Hodgkinson. To Hay 1816. To Eardisley 1818–1860. Tramroads were not always the ultimate answer; in 1845 at Llanvillo, 7 miles from Brecon: 'There are some quarries of good limestone and of stone for building and roofing houses, the working of which affords employment to a few persons. The Hay tram-road passes within a mile of the place but the road leading to it is at present in so bad a state that very little benefit is derived from it.'
Hereford	From Junc. w. Grosmont Rly at Llangua Bridge to Wye Bridge, Hereford (11¾ m.); J. Hodgkinson, 1829–*c.*1852. Under the Act, 26th May, 1826, locomotive engines were allowed and toll rates quoted for all manner of things, dung being 2*d.* (1p) per ton per mile; coals, iron, slates, etc. 4*d.* (1½p) per ton per mile. Cows were 2*d.* each, sheep and swine 1*s.* 3*d.* (6p) per score and with some foresight: 'For all Waggons and Carriages carrying Persons for Hire on the said Railway, for each Person so carried . . . 2*d.* per mile'.[3] More modern than the Llanfihangel below, 4 ft rails were used of a similar section, resting on 4 ft 4 in. long cast iron sleepers each weighing 53 lb., the rails being retained by troughs on each end with a wooden wedge to give some flexibility. These sleepers in turn rested on, but were not attached to, stone blocks. Chippings both held the track *in situ* and provided a good grip for horses' hooves. On the opening day, 21st September, 1829, 33 train loads of coal arrived, 15 from Blaenavon and 18 from Pontypool.
Hills	From Llanfoist to Garndyrys Ironworks (5 m.); T. Hill, 1825–1853. But see also Pwll-Du, the line from Blaenavon to Llanfoist being commonly called 'Mr. Hill's Tramroad'.
Llammarch	From Llammarch Colliery to Clydach Ironworks (1⅜ m.); T. Dadford Jnr, 1795–1860. Extn to Gilwern (2½ m.); H. Henshall, 1803–1860.
Llanfihangel	From Govilon to Llanfihangel Crucorney (6½ m.); W. Crossley, 1814–*c.*1852. As with the Grosmont, this tramroad consisted of cast iron plate rails, 3 ft long, 4 in. wide on the tread and with a vertical flange 3 in. in depth laid on cubical stone blocks, each weighing about a long cwt. (120 lb.). 4 in. spikes were driven through a hole in the stone. Gauge of 3 ft 6 in. to the inside of the flanges.
Llangattock	From Llangattock to Daren Cilau Quarries (2 m.); J. Bailey, 1816–1911. Later extended to Nant-y-Glo (4 m.); 1825–1911.
Pwll-Du	From Blaenavon via tunnel through coal mine (realignment of Tyla Line) to Pwll-Du (1¾ m.); NK, *c.*1815–*c.*1854. Extended to Garndyrys Ironworks, Junc. w. Hills Trd. 1817.
Tyla Quarry	From Blaenavon Old Works to Pwll-Du and Tyla (2 m.); NK, *c.*1789–*c.*1815.

TRAMS AROUND THE BLORENGE 1845

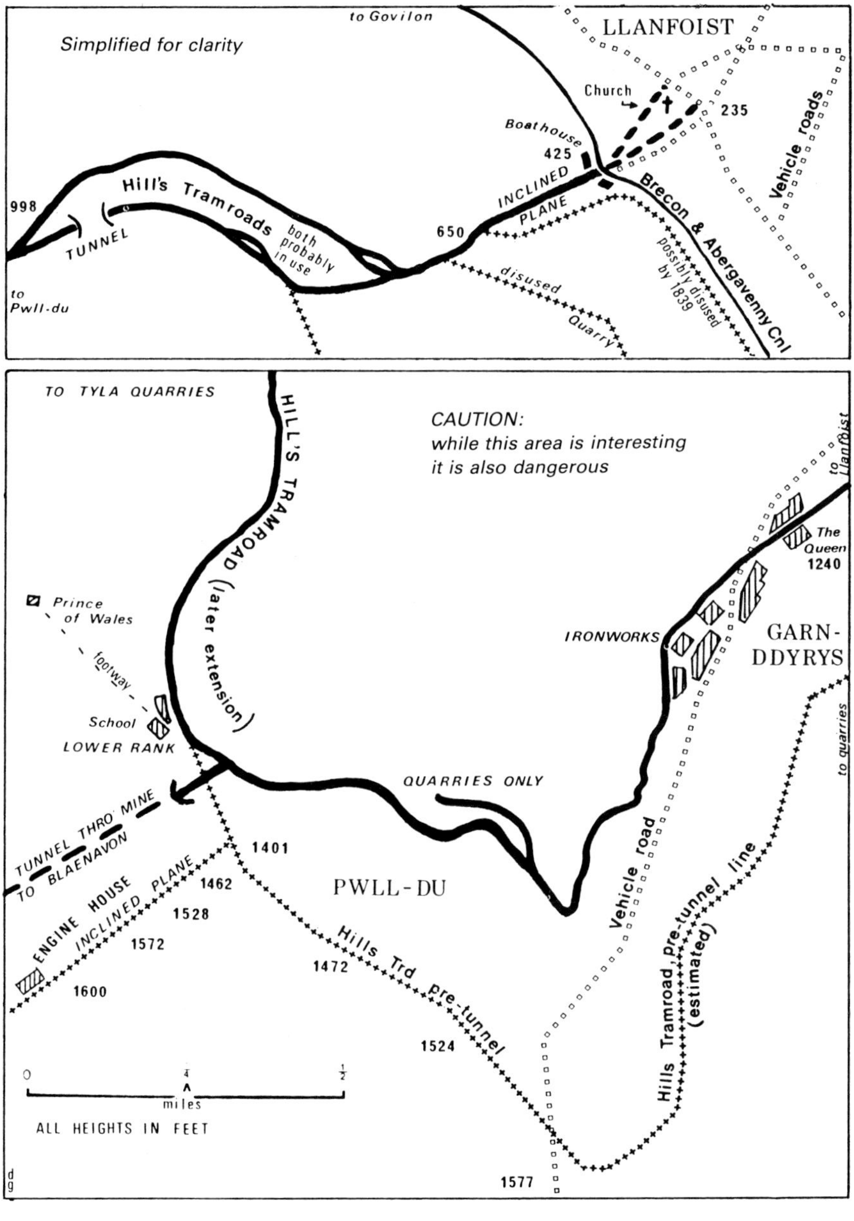

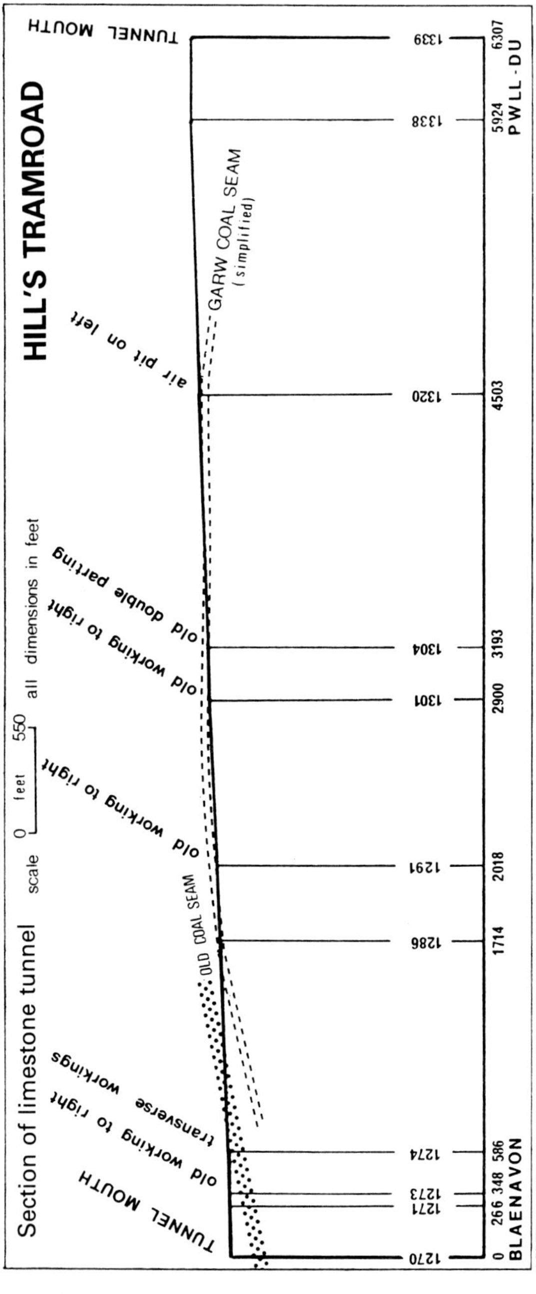
HILL'S TRAMROAD
Section of limestone tunnel
scale 0 feet 550
all dimensions in feet
TUNNEL MOUTH
old working to right
transverse workings
OLD COAL SEAM
old working to right
old working to right
old double parting
air pit on left
GARW COAL SEAM
(simplified)
TUNNEL MOUTH
1270 1271 1273 1274 1286 1291 1301 1304 1320 1338 1339
0 266 348 586 1714 2018 2900 3193 4503 5924 6307
BLAENAVON
PWLL-DU

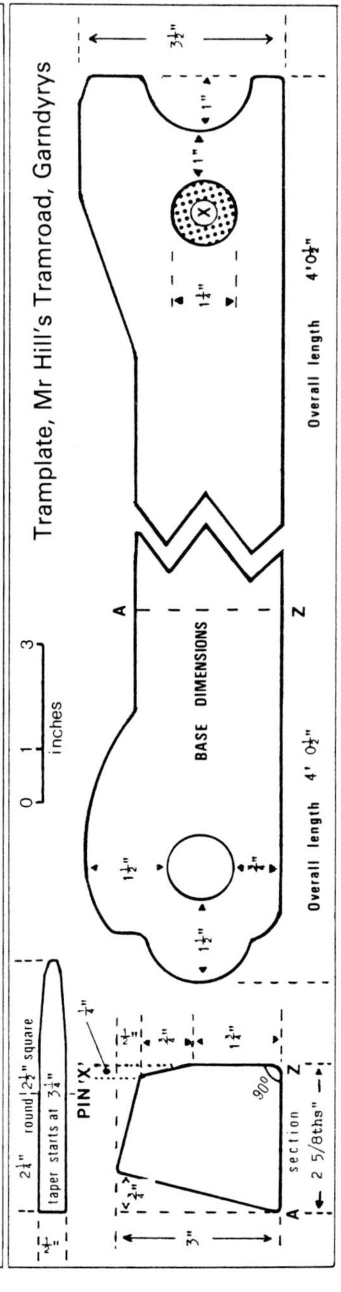
Tramplate, Mr Hill's Tramroad, Garndyrys
0 1 3
inches
BASE DIMENSIONS
A
Z
Overall length 4' 0½"
Overall length 4'0½"
PIN 'X'
taper starts at 3¼"
2¼" round 2½" square
section
2 5/8ths"
90°
3"
3½"

The Doctor's Canal:

Hafod	From Treforest to Hafod Colliery (3¼ m.); NK, 1809–1914.
Dinas Uchaf	From Junc. w. Hafod Trd, near Gyfeillon to Coffin's Colliery, Dinas (2¼ m); D. Naunton, 1812–1914. 'A railroad passes from the Cardiff canal up the right bank of the Rhondda to some coal-pits in the upper part of the vale and another railway has been constructed . . . to Dinas coalworks, whence 300 to 400 tons of coal are conveyed daily to the port of Cardiff by an engine . . .'[4].

Glamorganshire Canal

Castle Morlais	From Junc. w. Dowlais Trd, Penydarren to Castle Morlais Quarries (2 m.); R. Hill. 1803–? 'It is agreed and determined to lay a Tram Road of the width of the Road already laid (4 ft 2 in.) . . . the present Road to remain on the inside of the Wide Road, and kept in repair. R. Hill is to be allow'd £150 for making and laying . . . The whole road to be completed by the 25th December, 1803, and what plates each party may have furnished more than their proportion to be paid for the same by the Treasurer at six pounds sixteen shillings and 6*d*. p. ton Long Wt. On the 25th December, 1803, the Accnts to be settled and the Capital of each party adjusted, after which the Repairs of the Road and all Expences to be paid from the Tonnage arising upon the Road, Provided the Tonnage is sufficient; if not, to be paid in the proportions as hitherto by each party'.[2]
Dowlais	From Jackson's Bridge, Merthyr to Dowlais Ironworks (1½ m.); G. Overton, 1793–1853.
Maes-mawr	From Upper Boat to Dihewydd Collieries (2¼ m.); of Lantwit Fardre on the line of this tramroad, it was said in 1845: 'The river Taf runs along the eastern confines of this parish which abounds with coal, worked for exportation at the Maes-mawr collieries, where about forty persons are employed: from these works a tram-road communicates with the Glamorganshire canal . . . In the North eastern part of the parish . . . a populous and increasing village has been brought into existence by the erection of very extensive works for the manufacture of patent wrought-iron railway plates; and many cottages have also been built by Mr. Crawshay. The river here is crossed by two bridges, over one of which is a tram-road, connecting the new works with the canal, which passes within a few yards on the other side of the river.'[4]
Pen-tyrch	From Melingriffith Works to Pen-tyrch Forge (2½ m.); NK, c.1815–c.1877.
Penydarren	From Abercynon to Penydarren Ironworks (9½ m.); G. Overton, 1802–1880. It was upon this tramroad that on 21st February, 1804, as a result of a wager of 500 guineas between Samuel Homfray and Anthony Hill, Trevithick's locomotive engine was run. Although successful in that it completed the journey from Penydarren to Abercynon, a distance of 9½ miles, dragging a load of 10 ton and 70 men in or on 5 waggons, Homfray winning his bet, nevertheless due to the unbalanced reciprocating masses involved and the resulting breakage of the tramplates, the matter was not pursued with any enthusiasm thereafter.
Powells	From Abercynon to Gelligaer Colliery (2¾ m.).

The old iron tramway bridge across the river Cynon at Robertstown, Aberdare. Cast at the Abernant iron works, it is dated 1811. *G. Davies*

Penydarren Tramroad, Glamorganshire Canal and Penydarren Ironworks, *c*.1811. *Waterways Research Centre*

Kidwelly & Llanelly Canal

Cwm Mawr	From Cwm Mawr to Pontyberem (1¼ m.); NK, c.1840–? Replaced incomplete canal line for colliery use.
New Lodge	From Burry Port to New Lodge and Parc-y-Goch (1–2½ m.); James Green, c.1832–c.1886. Various branches.
Trimsaran	From Moat Farm to Trimsaran Collieries (1 m.); NK, 1816–c.1886.

Monmouthshire Canal

Abercarn Rly	From Cross Keys via Hafod Trisliog Colliery to Cwmcarn; (7 m.); B. Hall, c.1805/1814–1877. Various branches.
Abersychan	From Blaenavon Trd, Abersychan to Abersychan Ironworks (1¼ m.); NK, 1827–1860.
Blaenavon	From Pontnewynydd to Blaenavon Ironworks (5¼ m.); T. Dadford Jnr, 1796–1853. Rises 610 ft in this distance.
Blaendare	From Pontymoile to Blaendare Ironworks (1 m.); T. Dadford Jnr, 1796–c.1818. Extn to Balen-y-Cywm (4 m.).
Caerleon	From Halfway House, Cwmbran to Caerleon (4 m.); NK, c.1799–c.1865.
Ebbw Vale	From Crumlin Bridge to Beaufort Ironworks (9½ m.); G. Boye, 1796–1860. In 1800 Benjamin Outram was called upon to report on various tramroads in the area and said: 'The railways to Beaufort and Blaenavon are better laid out than most I have seen on that system. The Trosnant road should have been constructed on a more gradual gradient, but they are all constructed on a principle far inferior to the system I have and which after several years proof in a great variety of situations has been found to exceed every other make hitherto adopted. The labour of one horse on a railway of this kind being equal to the labour of four horses on the best railways that have been constructed on the plan of those belonging to the Monmouthshire Canal Company.' He recommended flanged plates on stone blocks, his suggestions being carried out, but on 17th March,1802 one trader was most upset: 'In consequence of your having suffered the old rail road to be cut to pieces and rendered impassable we have at considerable expense altered our wagons to the principle of your new road, but we find that it is so badly made that we have not been able to haul upon it.' Incidentally, by 1806 regulations restricted the total weight of load and wagon to 56 cwt. following the breakage of many tram plates.
Harfords	From Ebbw Vale Rolling Mills to Junc. w. Sirhowy Trd (? m.); NK, 1818–1839.
Llanhilleth	From Crumlin Bridge to Cwm Cyffin via Llanhilleth Collieries (2 m.); NK, c.1799–c.1819.
Nant-y-Glo	From Ebbw Vale Trd, Aberbeeg to Coalbrook Vale Ironworks and Nant-y-Glo (5½ m.); W. Wells, 1824–1860.
Park Mile	From Nine Mile Point Line (1½ m. from Newport terminus) through private park around Tredegar house (1 m.); J. Hodgkinson, 1824–closed by 1855.
Pen-llwyn	From Sirhowy Trd, Nine Mile Point to Blackwood (5 m.); J. Hodgkinson, 1824–closed by 1855. Nine Mile Point to Ynys-ddu vested in Sirhowy Tramroad Company, renamed Llanarth Tramroad.

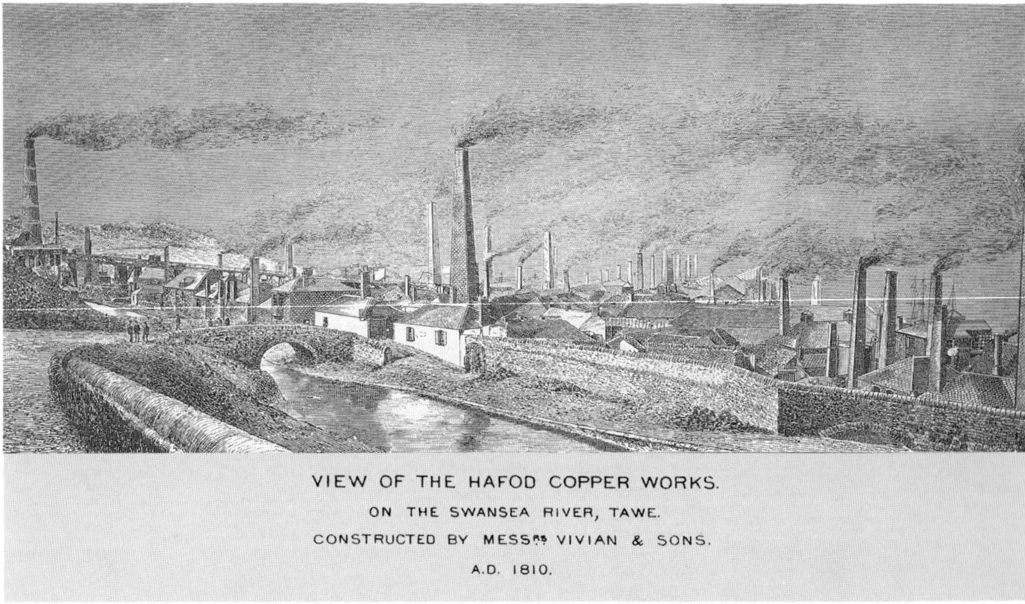

Vivian & Sons, Hafod Copper Works as seen in 1810. This is one of the few illustrations showing the Llansamlet or Smiths canal. A decade later a parallel tramroad was built simply because the waterway was unable to cope with the traffic on offer.

The Varteg Tramroad incline is over half a mile in length; seen in 1973.

RAILS AND PLATES TO VARTEG 1875

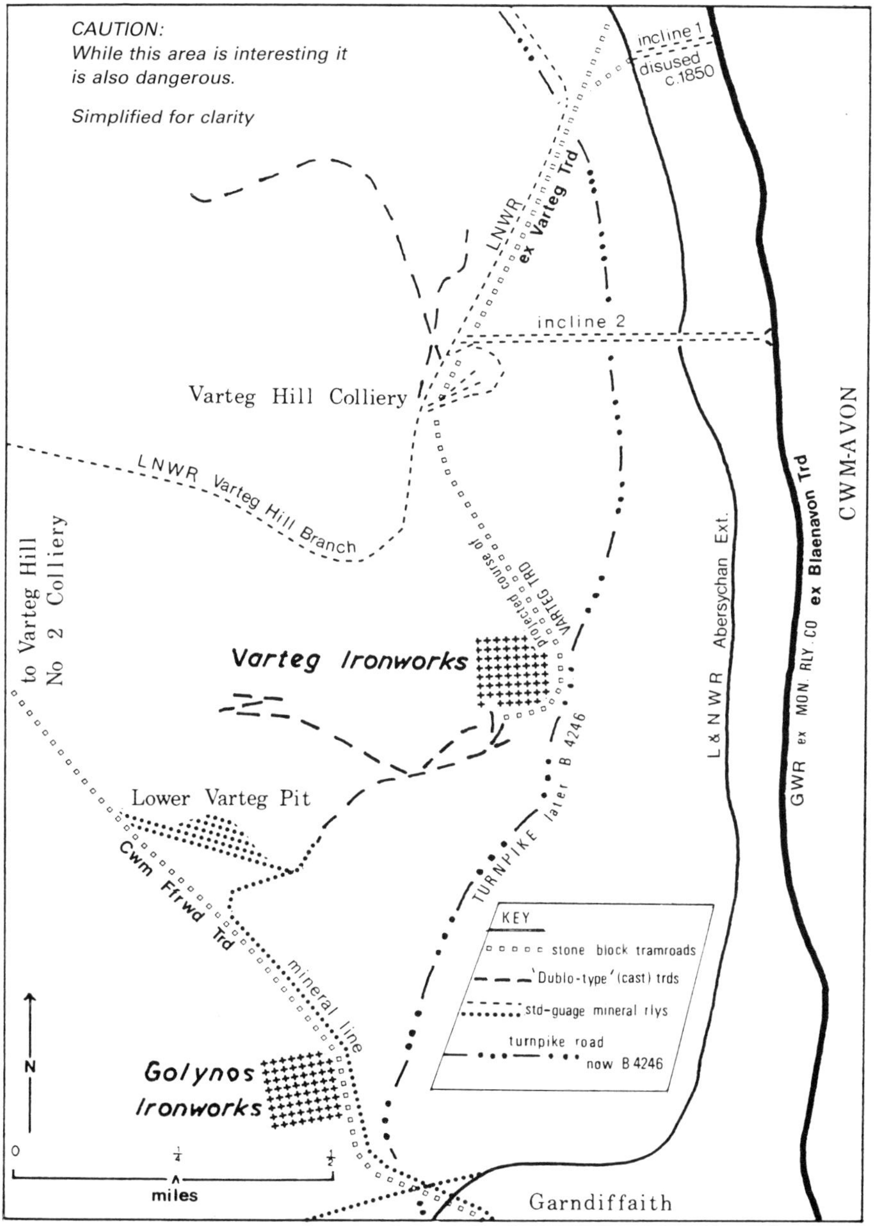

Pontypool	From Blaenavon Trd, Ponynewynydd to Trosnant Furnace, Pontypool (¾ m.); J. Hodgkinson, 1829–1860.
Porthmawr	From Halfway House, Cwmbran to Mynydd Maen Collieries (2 m.); NK, c.1799–1900.
Rassa	From Sirhowy Ironworks to Beaufort Ironworks and Junc. w. Ebbw Vale Trd (2½ m.); T. Dadford Jnr, 1797–1860.
Rumney	From Nine Mile Point Trd, Bassaleg to Rhymney Ironworks (21¾ m.); G. Overton, 1836–1861. Act 20th May, 1825. Subscribers were Sir Charles Morgan, Joseph Bailey, William Thompson and Crawshay Bailey, who put up the whole of the estimated £47,850. Having a rise of 756 ft, 3 inclined planes were required, 'the first of which, commencing at Pye Corner, is two miles and three furlongs in length, and rises 114 feet; being but one-third of an inch in the yard; the next plane is ten miles, six furlongs and eight chains in length to Gellyhave Colliery, and rises 209 feet or one-eighth of an inch per yard; the remainder of the railway is eight miles, four furlongs and two chains in length, and rises 433 feet or one-third of an inch per yard.'[3]
Nine Mile Point	From Newport to Nine Mile Point (9 m.); J. Hodgkinson, 1824–closed by 1855.
Sirhowy	Continuation of Nine Mile Point Trd to Sirhowy Ironworks (16 m.); J. Hodgkinson, 1805–closed by 1855.

The above three tramroads were collectively known as the Sirhowy Tramroad. 'This rail-way was completed about twenty years ago, also a turn-pike road by the side of it for about 17 miles, the total expense amounting to about £74,000 or about £3,000 per mile. About £40,000 of this sum was expended by the canal company in consequence of building a bridge, and some very deep and expensive cutting, whilst the Tredegar iron company completed nearly double the distance at the cost of £30,000 . . . Notwithstanding the expense, this road pays the proprietors 30 per cent, having a considerable trade upon it in coal and iron . . . The coal and iron are conveyed upon it in waggons, each carrying about forty-five to fifty hundredweight, exclusive of the waggon; and a team of four or five horses will draw about fifteen of these waggons down, and take the same number of empty ones back, with ease . . . This is a noble monument to the spirit and industry of Britons'.[21] The first regular passenger service on any of the Valleys tramroads was operated from 1822 between Newport and Tredegar at a toll of 6*d*. (2½p) per vehicle at each toll gate although this ceased some time before 1850.

Tredegar	From Tredegar Ironworks to Trevil Quarries (3 m.); T. Dadford Jnr, 1804–1860. Branch to Union Ironworks, Rhymney (2½ m.).
Trevil	From Trevil Quarries to Junc. w. Rassa Trd at Trevil Machine (2⅝ m.); T. Dadford Jnr, 1797–1860, and from Junc. w. Rassa Trd at Shop Row, Rhyd-y-Blew to Pont-y-Gof (¾ m.).
Trosnant	From Pontypool to Old Furnace (1 m.); T. Dadford Jnr, 1796–1860. Extn to Pontymoile (½ m.); J. Hodgkinson, 1810.
Varteg	From Blaenavon Trd, Cwmavon to No. 1 Colliery, Varteg Hill, later Varteg Ironworks (2½ m.); NK, 1819, rebuilt 1837–1860

TRAMPLATES 1805–1850

Not to scale

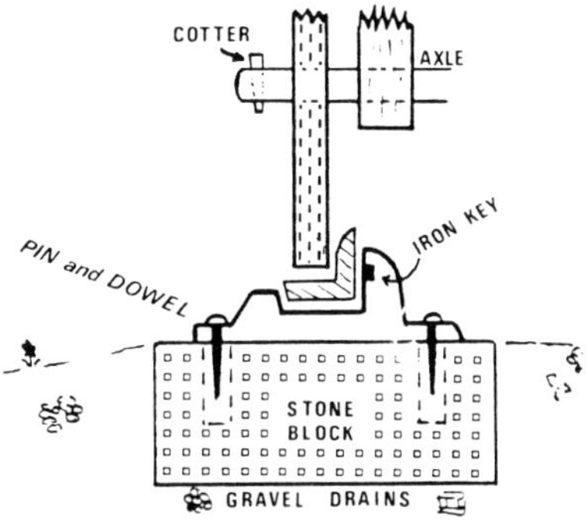

Sirhowy 1805

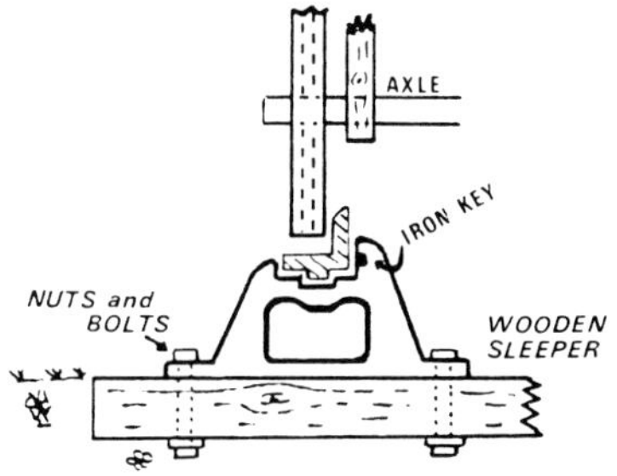

Sirhowy 1850

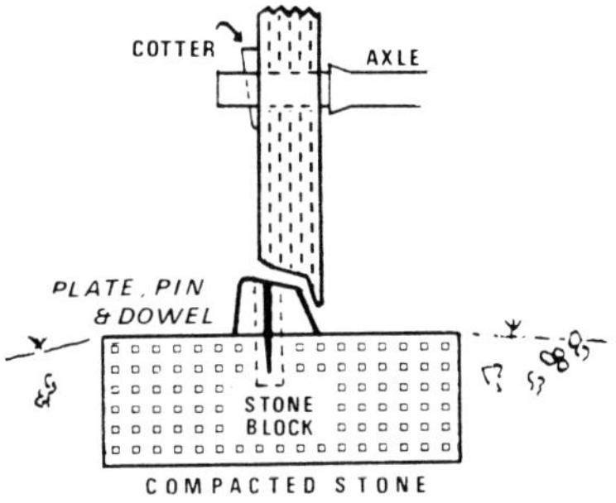

Hill's 1840

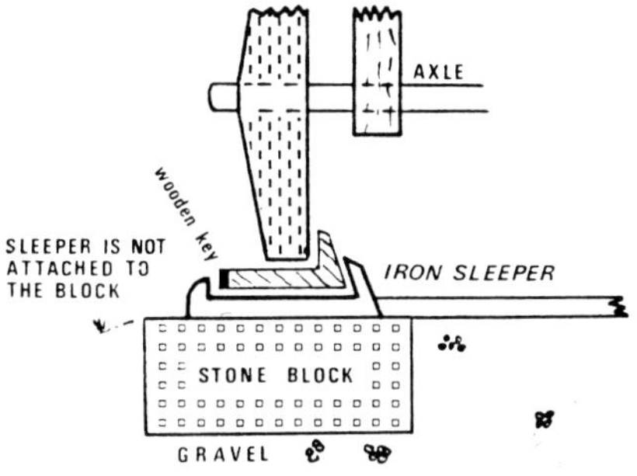

Hereford 1829

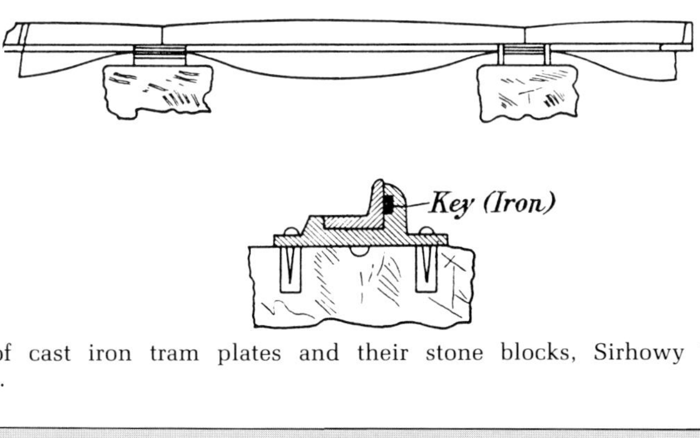

Sketches of cast iron tram plates and their stone blocks, Sirhowy Tramway 1805–1840.

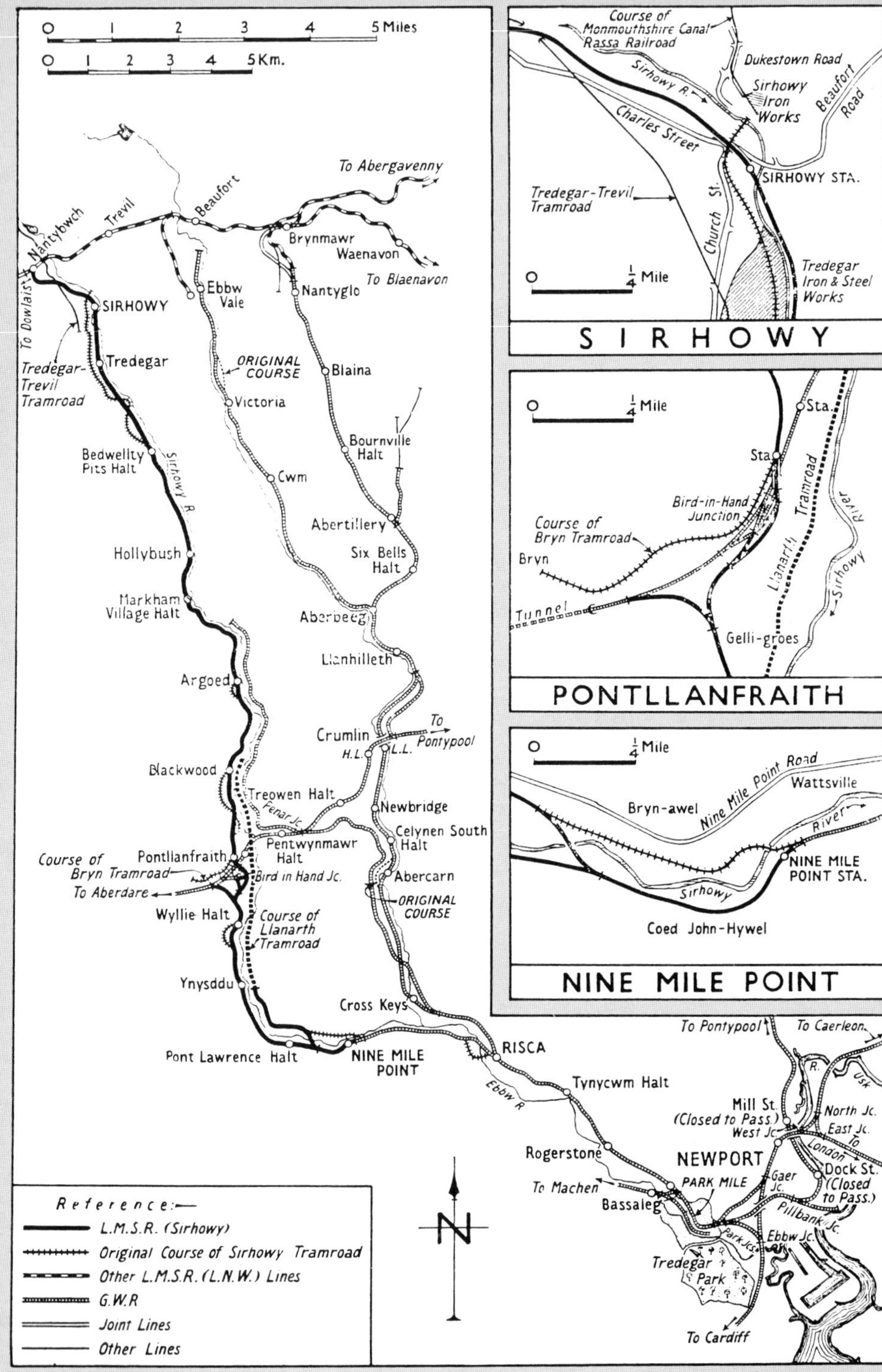

The Sirhowy Tramroad, the connecting tramroads and later railway lines are shown in this illustration of 1940; many sections can still be traced on modern Ordnance Survey maps.
Courtesy Railway Magazine

Neath Canal:

Abernant — From Glyn-Neath to Abernant (8 m.); E. Hopkin, 1805–c.1860 'Since . . . 26th February, 1802, Messrs Homfray and Birch . . . have gone into divers contracts . . . opened a railroad from furnace to Neath Canal with incline plane and steam engine to work same, and in executing same have spent £70,000 the money of Messrs Tappenden, said Homfray and Birch having not paid anything . . . in consideration of 10s. paid Homfray and Birch release to Messrs Tappenden all shares in tramroad from Abernant Furnaces to Neath Canal, tollhouse, bridge, culverts, etc. and lands consigned to them.'[22]

Aberpergwm — From Aberpergwm to Pwllfa'r-on (1 m.); NK, 1814–c.1876. A 1 m. long inclined plane.

Dr Bevan's — From Pontwalby to Dinas Quarries (1½ m.); T. Cartwright, 1807–1925. 'Dr Bevan of Neath, is now engaged at his own expence, in making a Tram-Road from Dinas Rock, in the county of Brecon, to communicate with the Neath Canal, passing through the estates of the most noble the Marquis of Bute and Walter Wilkins, Esq., M.P.through the parish of Ystradavodwg, in the county of Glamorgan, being upwards of one mile and a half long. The whole line is already cut and formed, the tram-plates are now laying down, under the direction of that able surveyor Mr Evan Hopkin and the under-taking is in such a state of forwardness, that the whole will be completed by the first week, in July next. This Tram-Road will be of the greatest public utility and emolument, not only to the Proprietors of the Neath Canal Navigation, but also to the neighbourhood in general, as the farmers will thereby have that important article limestone delivered on the upper bank of the Neath Canal at a very cheap rate. Dinas is an immense rock consisting of limestone, and excellent fire-clay much superior to that of Stourbridge or Flintshire; a black clay sometimes used by potters; black and variegated marble, terra ponderosa, iron ore, and lead ore; the joint property of the beforementioned Dr Bevan and John Bevan Esq. Many of these articles are also found on the estate of the Marquis of Bute, near the same spot, and adjoining there is limestone on the estate of the Right Hon. Lord Vernon, and coal and iron ore on the estate of Walter Wilkins, Esq. contiguous thereto. This Tram-Road will afford a cheap and easy conveyance for all these valuable articles to the Neath Canal, as it will be open to all, on paying their proportion of tonnage.'[23] In the same newspaper for 2nd May, 1807 appears the following advertisement: 'Persons willing to Contract for the Conveyance of 30,000 tons of Fire-Clay (ready raised and the drams filled) from Dinas Rock, situate in the parish of Penderin, in the county of Brecon, to the upper end of the Neath Canal, in the parish of Cadoxton, near Neath, in the county of Glamorgan, to be carried along the Dram-Road belonging to Dr Bevan, of Neath, and discharged on the Canal Bank, between the 1st of May 1807 and the 1st of May 1808, are desired to send proposals stating the lowest price per ton for which they will undertake the same to Dr Bevan aforesaid, None but principals will be treated with'. When, some

The Waun Lwyd Tramroad seen on its ascent of the mountain (Cefn Manmoel, Ebbwvale), 1973.

The Llangattock tramroad is for a large part of the way easy to follow with some tramplates and artefacts surprisingly easy to find.

	40 years later, a tram went off the rails near Dinas Rock, Evan Bevan, the local bard, composed the following triban to 'plague' the haulier, Thomas Jones: *Mae'n Hynod Yma Heno* *Mae'r Dramrod Wedi Digio* *Fe Aeth Y Ddram I Lawr I'r Cwm* *Pan Ffaelodd Twm Ag Halio!* Poetically translated by E.G. Millward as: *Things are very strange tonight,* *The tramroads in a sorry plight,* *The tram pitched down into the vale* *And Twm the haulier turned quite pale!*
Clyne	From Ynys Nedd Farm to Cwm blaen Palena (2 m.); NK, *c*.1800–1867. Branch to Glyncorrwg (2 m.).
Parsons Folly	From Pant-y-Coed to Blaen-Cregan Collieries (5¾ m.); E. Parsons, *c*.1847–1867.
Pont Nedd Fechan	From Cefn Rhigos Trd, Glyn Neath to Pwll-Du Ar Byrrdin (1¾ m.); E. Hopkins, 1812–*c*.1860.

Smith's Canal

Old Waggon Way	From Llansamlet Along River Tawe (2½ m.); Chauncey Townshend, *c*.1743–*c*.1784.
Scott Tramway	From Foxhole to Scott's Colliery, Llansamlet (1 m.); NK, 1816–*c*.1860. (Relics of 'Cornish' engine-house now preserved at Colliery site.) Probably later truncated to terminate at Canal Head.

Swansea Canal

Brecon Forest	From Sennybridge to Drum Colliery (12 m.); J. Christie, 1825–1859.
Clayponds	From Ynysgedwyn to Junc. w. Brecon Forest Trd at Onllwyn (4 m.); J. & R. Brunton, 1834–1867.
Clydach	From Clydach-on-Tawe to Craig Merthyr Collieries (2¾ m.).
Cribarth	From Pen-tawe to High Level Quarries, Cribarth (2½ m.).
Hen-Neuadd	From Pen-tawe to Cribarth Quarries (1¾ m.); T. Sheasby, 1798–?
Oystermouth	From Brewery Bank to Castle Hill (7½ m.); E. Martin, 1806–1877. Act 29th June, 1804.
Palleg	From Ynysgedwyn to Palleg Colliery (1½ m.); NK, 1807–*c*.1850.
Waun-y-coed	From Ynysmeudwy to Waun-y-coed Colliery (1¼ m.); Branch to Cwm Nant Llwyd Colliery (¾ m.); 1801–*c*.1900.

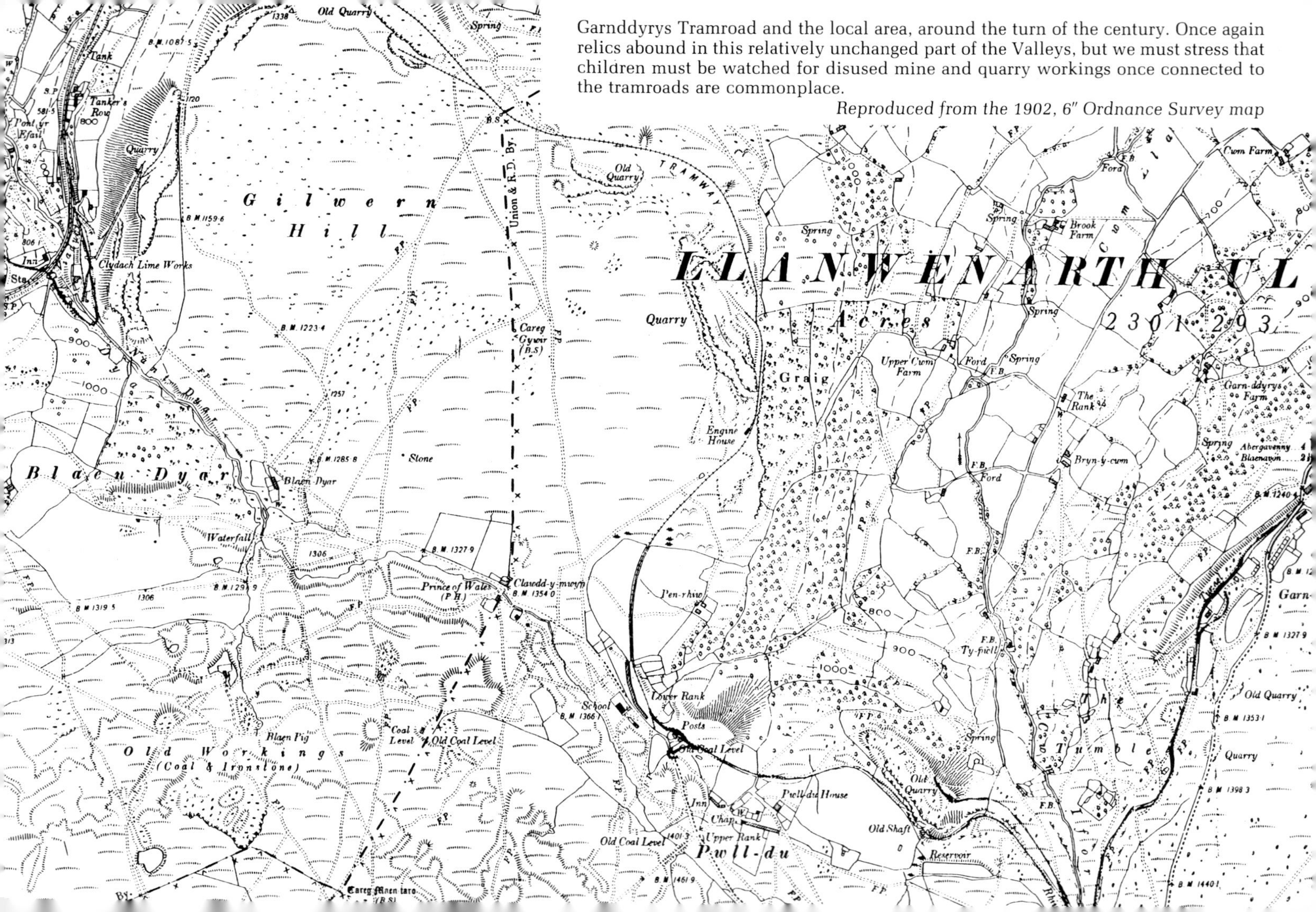

Garnddyrys Tramroad and the local area, around the turn of the century. Once again relics abound in this relatively unchanged part of the Valleys, but we must stress that children must be watched for disused mine and quarry workings once connected to the tramroads are commonplace.

Reproduced from the 1902, 6″ Ordnance Survey map

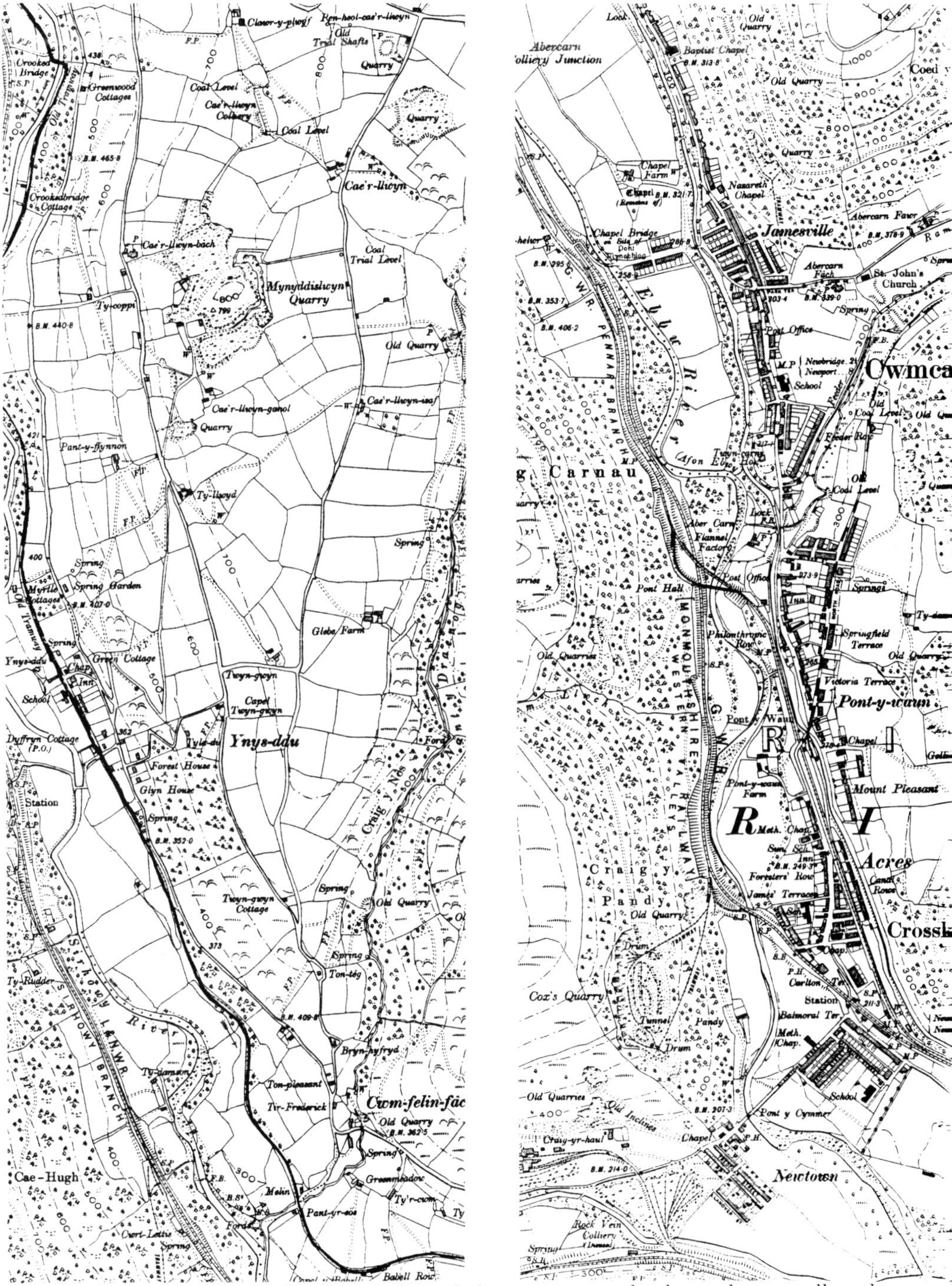

Cwmcarn Colliery closed recently and the area is now landscaped as a walk and camping site. The canal is sad, but the vicinity justifies a day searching for relics. The tramroad at Ynys-ddu is easily found and an interesting example of a one-time tramway. *Reproduced from the 1902, 6″ Ordnance Survey map*

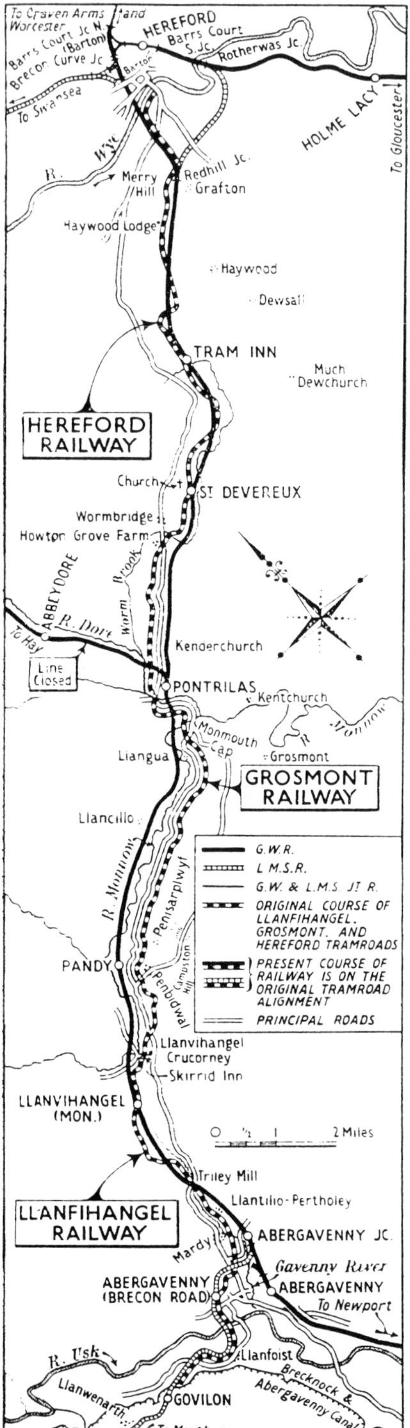

Courtesy Railway Magazine

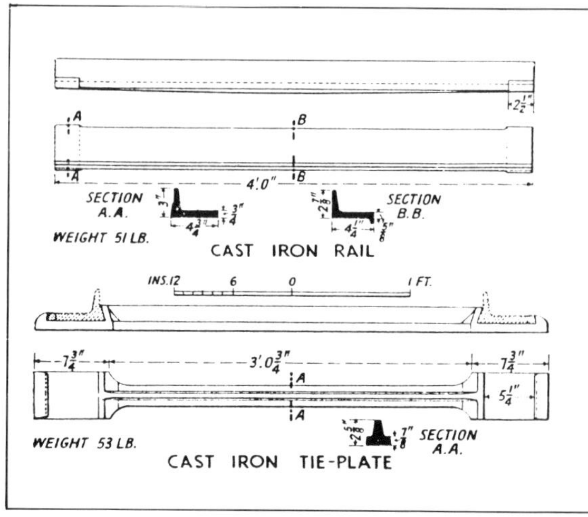

The cast iron tramplates and rails developed by John Hodgkinson for use on the Hereford Railway 1829 and still visible in the 1950s.

Restored lock on the Neath Canal at Resolven.

Chapter Four

Tramroad Relics

Searching for relics of tramroads can be both interesting, inasmuch as one is led to ironworks, canal wharves and collieries, and pleasant, as the family can accompany one without difficulty. Obviously it is impossible in this book to give the details of all tramroads, but with the aid of an Ordnance Survey 1 in. to the mile map (or, better, 6 in. or 25 in. – your library may have copies) the basis of a trip can easily be planned. Some parts of the route may be through private land; if you go through gates, close them! In the Merthyr Vale some hostility may be found; too many ghouls went to gape at Aberfan. Conversely in and near Blaenavon the Welsh are jovial and almost excessively helpful; a most knowledgeable man near the ironworks is also the local falconer. On the B 4246, Blaenavon to Varteg is a superb incline, details of which – at least its usage in latter days – were given to us by the local poacher!

The following notes, based on those of Geoff Smith in the Monmouthshire Railway Society's Journal, may serve as a guide:

1. *Llangattock Tramroad.* Initially a plate-way and latterly an edge-way, the route from Darren Cilau to Llangattock was afterwards extended south and west to the Nant-y-Glo furnaces of J. & C. Bailey.

Start at Crickhowell (SO 218185) take the Gilwern road (A 4077) which turns sharp left at the foot of the Usk river bridge, turn right through to Llangattock (SO 209174), shortly beyond this is the canal bridge from which, in the Brecon direction, can be seen a sizeable complex of lime kilns while beyond Bridge No. 115 on the off-side of the canal, are the original lime kilns and wharves of 1816.

In the grass verge for the first hundred yards of the roadside will be seen GWR boundary posts after which on your left is a lane well studded with stone blocks from the tramroad. Shortly the lane opens out alongside a minor road revealing two track beds, the later one went parallel to the road through to the kilns. At the junction of the track beds are the remains of a weighbridge and soon after road and tramroad part company, the tramroad heading for the hills along the right hand side of a wide path, a distance of one mile. The farmer does not object to walkers provided they close the gates.

After the mile the tramroad climbs by a double track incline to the remains of a winding house. This is followed by a second incline running diagonally up the hillside. Much of the head work of the incline is in evidence at SO 201159 including the stone guides which led the rope from the top drum to a well where it was attached to a bucket; by filling the bucket with water a system of pulleys enabled the wagons to be drawn up as the bucket went down. The wagons reaching the top, doors in the bottom of the bucket were opened and the water discharged. A descending load would return the now empty bucket to the top.

The main tramroad followed the track to the south-east with branches leading to the quarry faces above; exploration is dangerous due to unstable ground. After ⅔-mile the tramroad finished under the Brynmawr road.

Thereafter odd bits and pieces may still be found, especially the Wern Watkin incline (SO 207156) and *en route* to Brynmawr the line of the LNWR is visible, a latter-day substitute for Hill's tramroad.

2. *The Hereford Railway*. Nicknamed the 'lazy man's tramroad', it comprised three separate undertakings, the Llanfihangel, Grosmont and Hereford railways. In 1846 the three lines were purchased by the Newport, Hereford & Abergavenny Railway, who built their orthodox standard gauge railway over some parts, while from Abergavenny to Llanfoist it more or less disappeared under the track of an LNWR satellite, the Merthyr, Tredegar & Abergavenny Railway.

It is not impossible that these tramroads, together with others, make up the longest chain in the world. From Hereford the plates extended to Abergavenny and then to Govilon. At Govilon, Bailey's tramroad went via Brynmawr to Nant-y-Glo, giving a connection to the Monmouthshire Canal Company's line through to Crumlin. Alternatively, if the proximity of Bailey's and the Beaufort tramroads in Brynmawr are accepted as a 'connection', an even longer road via the Beaufort, Tredegar, Sirhowy, Nine-Mile-Point and Risca tramroads would take one to Newport. Alas for dreams; numerous changes of gauge would make the through journey difficult, although not of course for the horse!

From Govilon wharf (SO 271137) to Llanfoist the ground is enclosed but odd blocks do exist here and there. From Llanfoist to Abergavenny the old Merthyr, Tredegar & Abergavenny Railway, itself now defunct, was built on the track of the tramroad. From the road (B 4521) by the erstwhile entrance gates to Abergavenny Junction station (SO 310153) a trace can be seen in the fields to the west across the river.

Return to the main (A 465) road, and head towards Hereford. At Mardy the first village, the trace joins the road on the east side, a distinct level can be seen in the fields just below the road for the next mile or so northwards. As Triley Bridge (SO 310172) nears, the course vanishes – presumably under the railway. What appears to be a trace exists to the east of the railway in the grounds of Triley Mill. Again, the trace vanishes in private land, and it is necessary to get to the Old Hereford Road to see more. Take the minor road that runs west from Triley Bridge, right at Ty Gwyn and right again at the T-junction that follows shortly. Continue up the Old Hereford Road for three miles, just past Llwyn-Gwyn (SO 308188) and over the hedge on the east side, the tramroad course is well defined about 200 yards away across the fields.

Tramroad and railway join again near the site of Llanfihangel station. Follow the minor road to rejoin the A 465 at Crossways (SO 323200) and turn along this towards Hereford. Just beyond the village of Llanfihangel Crucorny, park near the minor road leading to Llangattock Lingoed (SO 329210), for the footwork now begins. The road here is in a cutting, on the east side; half way down the hill a flight of steps leads up the side of the bank to the trace which is a footpath at the top. The path leads north-east on a substantial embankment where stone blocks abound.

Continue up the A 465 past Pandy. From there to Llangua the tramroad was constantly on the east side of the road. In the last ten years vast road

Hill's Tramroad, blocks and tramplates.

Hill's Tramroad, Llanfoist; tunnel under Brecon & Abergavenny Canal.

improvements have eliminated most of the tramroad but some fine bits remain bordering laybys. Maerdy (SO 371247) is a particularly fine example.

From Llangua through Pontrilas little exists to show where once the plates ran. About 1½ miles beyond Pontrilas, after the Sawmill, the trace reappears to the west of the road, sometimes as a wide verge sometimes just beyond the hedge, and stays more or less intact in that position for the next two miles. At about SO 422302 a deep cutting occurs. The main road is crossed at SO 425303 and the trace heads for St Devereux station on a tree lined bank, past the houses of Wormbridge.

Beyond, the railway course obliterates nearly everything . The next station, Tram Inn, commemorates the railway's ancestor. At Red Hill Junction the tramroad followed the Barton loop, the original railway course, running almost due north. The termination was a wharf on the Wye, under the new road bridge, approached by a fine curve on a slight embankment, easily seen from the bridge.

Many years disuse, a main line railway and a major trunk road have all failed to obliterate totally the Hereford Railway – surely a fitting tribute to the men who built it!

3. *Hill's Tramroad* is a fine example of an integrated transport system to serve the needs of one rather ramified organisation. The original tramroad ran from Blaenavon Works (SO 249093) via Garn-yr-Erw near Garreg Main Tare by way of inclines to Pwll-Du and on to the limestone quarries at Tyla. Garndyrys works were later built and in 1825 the tramroad was extended through the tunnel mouth via Garndyrys (SO 258120) to the Brecon & Abergavenny Canal at Llanfoist wharf. Thus limestone and coal came from the south end of the tunnel to Blaenavon Works while for Garndyrys, coal from the tunnel joined the limestone from Tyla, the finished product coming down the canal from Llanfoist.

This is a pedestrian's tramroad. On the B 4246 (Blaenavon–Abergavenny) road a group of cottages, once the 'Victoria' or 'Queen' Inn (SO 260122), have as their frontage the first part of the tramroad and the trace is a good path for the three or so miles to Llanfoist Wharf. On the way is a short tunnel and at the top of the inclines are remains of buildings, probably once the weighbridge and winding house. At Llanfoist Wharf the road may be rejoined after passing through a short tunnel under the canal and the tramroad may be seen bearing north from the wharf down to the village of Llanfoist. As the boatyard seemingly do not encourage casual visitors and access is rather poor it is not desireable to start exploration from Llanfoist Wharf.

The alternative way from Garndyrys is to start at the Victoria Inn where a short walk up the road brings one to a lane leading down to the furnace site from which the tramroad course can easily be seen. If you have come by car and left it at Garndyrys a long uphill struggle faces you as you retrace your steps; no doubt consulting the bus timetable beforehand will allow you a good day out with somebody else doing the driving!

From the furnace site, where the 'main line' of the tramroad ran in a tunnel under the furnace slag run, the trace can be seen leading SSW clinging to the hillside. A landslip has obliterated the course for a short distance, but the level can be seen on the other side of the broken ground.

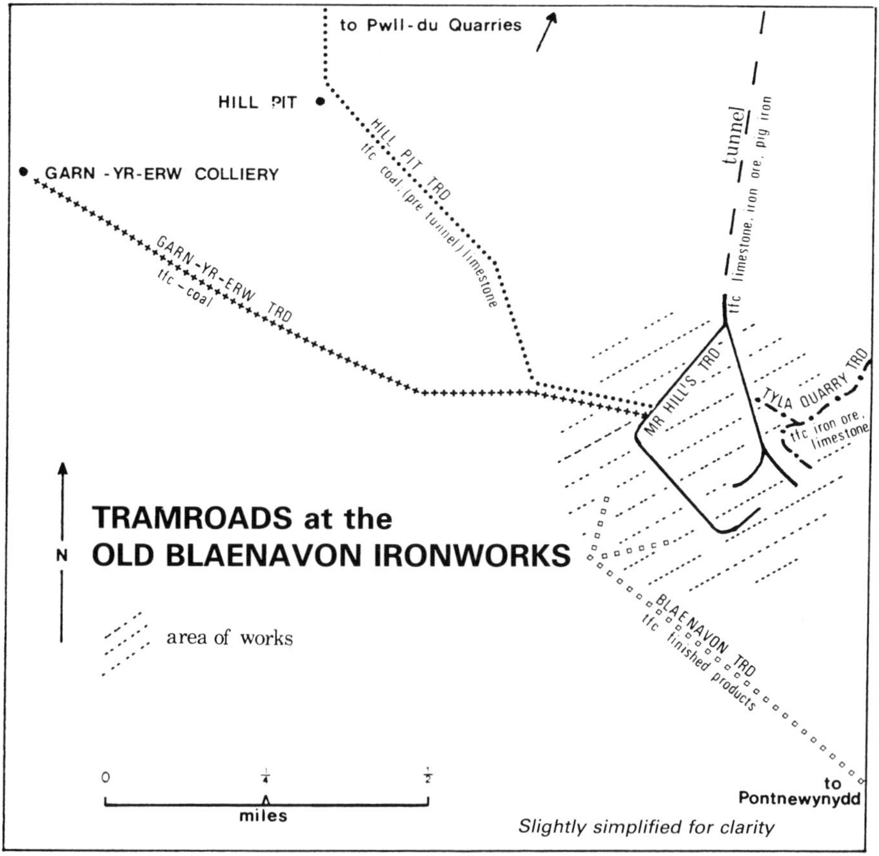
to Pwll-du Quarries
HILL PIT
HILL PIT TRD
tfc coal. (pre tunnel) limestone
GARN-YR-ERW COLLIERY
GARN-YR-ERW TRD
tfc – coal
tunnel
tfc limestone, iron ore, pig iron
MR HILL'S TRD
TYLA QUARRY TRD
tfc iron ore, limestone
TRAMROADS at the
OLD BLAENAVON IRONWORKS
N
area of works
BLAENAVON TRD
tfc finished products
0
1/4
1/2
miles
to
Pontnewynydd
Slightly simplified for clarity

3
TONS

Three-quarters of a mile further on the head of the valley is reached and the course turns abruptly NW. Just before this turn the remains of huts and a weighbridge can be seen on the left hand side of the tramroad. The tramroad can be seen on the opposite hillside, dividing into two traces, one climbing, one falling. The lower course leads to a lime quarry, and in summer is very heavily overgrown in parts. The upper course heads for Pwll-Du climbing steadily.

To find Pwll-Du leave Blaenavon by the Abergavenny road (B 4246), at the summit of the long climb from Blaenavon, a minor road diverges to the left by 'Keepers Pond' (SO 254107). Follow this road for about a mile, when the almost abandoned village of Pwll-Du is reached. The map shows houses, a pub, a chapel and a school, but now sheep are the bulk of the population. It must have been an awful place in winter, the north wind whistling up Cwm Llanwenarth and the mountains cutting off the low sun. Translated the name means 'Black Hole'!

The tramroad tunnel from Blaenavon emerges in Pwll-Du just below the road at SO 245117. The entrance is still clearly visible, but a fall blocks the way a few yards inside. Close by stood a row of cottages, 'Lower Rank', demolished in recent years. Some distance east at SO 246116, was another entrance to the tunnel, thus a triangular junction was formed between the two entrances, and a connecting line outside. The Garndyrys line ran away east whilst the Tyla road headed north-west in front of 'Lower Rank'.

Follow the track from the Lower Rank site, curving to the north. Shortly a wall is reached, then cross by the stile and the tramroad formation follows the hedge and is well defined. Another stile marks the return to open land. The trace can be made out, but following it is a wet business due to the marshy ground. After a hundred yards or so a road appears from the left and covers the tramroad course. This soon becomes a track and the extensive limestone quarries at Tyla ridge appear to the left. The track descends steadily as it passes the quarries and over the next two miles it curves steadily to the west, terminating almost above Clydach Village and pointing south, but various maps and authorities of various dates seem to disagree as to where the tramroad ended, and a narrow-gauge mineral railway serving the quarry began. Follow the track back towards Pwll-Du, at the end of the quarries leave the tramroad course to cross the bog and follow the tarmac road which is probably the original line that preceded the tunnel and crossed over the Carreg Maen Taro ridge by inclined planes, stone blocks with the appropriate holes show up through the potholes. Soon the Pwll-Du road is reached, the old tramroad following the old incline is at SO 248116 and can clearly be seen running straight up the hillside almost due south-west.

These bits and bobs of Hill's variegated tramroads should just be enough to whet your appetite; the relevant OS maps and numerous stone blocks show where many others ran.

4. *The Graig-yr-Hafod* was a rural rather than industrial tramroad whose sole purpose in life is alleged to have been the carriage of limestone from the slopes of the Blorenge to be burnt alongside the canal at Llanellen and

spread on the fields about. No doubt, however, there was coal and other necessities of a frugal existence to be carted up to the farms along the route.

Despite being a mere 3½ miles in length, one needs three maps to cover the entire course. Head along the A 4042 road between Pontypool and Abergavenny, pass Llanover then sharp left at SO 303102 about ½-mile before Llanellen village. Follow this lane up to the canal bridge (SO 301097), and, if you have come by car, continue until a wide verge on the right hand side leaves a convenient place to park.

Back at the canal bridge the remains of the lime-kilns are behind the house on the south side of the lane which climbs steeply; the tramroad here was in the field to the south crossing the lane where the gradient flattens, the wide verge on the north side pays homage to the rails that once ran there. We lose the course behind the Sunday Schoolhouse a few hundred yards further on.

At the top of the rise the tramroad crosses again from the north to the south of the lane and drops down onto its own course a little below. Gradually it climbs back to cross again near the rough track that comes down from the north, and a small quarry. Recent roadworks have eaten part of the route away, but it is soon picked up striking away from the road in a north-westerly direction. The bed, at first well defined, becomes progressively overgrown; this section is deep in the woods, almost forest, until abruptly intersected by the wall bounding Bryn-Bedw. Beyond, its course can be seen but should not be walked without permission.

Follow the wall down among the trees until the road is reached again, turn right and up the hill to where the track from Bryn-Bedw joins from the right (SO 289101). On the left is a gap in the hedge and a gate; the tramroad can be seen clearly along the hillside here skirting the house 'Bryn Deri'. Through the gate, the first hundred yards or so is heavy going through near jungle; the course follows the 'Bryn Deri' boundary hedge and in places has fallen down the steep hillside towards the river. Past 'Bryn Deri' however the ground is more stable and stone blocks mark the course abundantly. A deep (in tramroad terms) cutting appears, the valley widens and the farmhouse 'Ty-Gwyn' appears. The tramroad crosses the road before reaching the farm but is lost for a short distance beyond.

Pick up the course again, as the track to the north of 'Ty-Gwyn' (SO 284104) and where the path sign-posted 'Blorenge' diverges to the right the tramroad appears to fall steeply ahead. At first the track-bed is well defined, but gradually it deteriorates until intersected by a thicket and barbed wire fence; beyond is open land. The course is difficult to locate, but can be found as a ledge on the steep hillside. The course turns south-west, gradually climbing, until at approximately SO 257107 it meets the track that descends from the Blorenge; the track continues to fall towards a stream while the tramroad, almost imperceptibly, peels off to the right and continues to climb up the valley side, leading west. A patch of marshy ground and the course bends sharply left to head south-east. The 'bridge', a simple stone slab over the stream is still there. Still climbing, the course begins to turn south and re-enters private land, so permission should be sought before continuing.

Having turned fully south, the course flattens and follows the contours, ahead is the farm 'Craig-yr-Hafod' to the west of which the course passes into the quarries.

We wish you luck!

The canal head at Aberdare showing the sturdy crane on the wharfside, c.1930.
Courtesy Aberdare Library

Ashford tunnel on the Brecon & Abergavenny canal was unavoidably closed to navigation in 1984 and not re-opened until 5th May, 1985. The tunnel had become mis-shapen due primarily to traffic on the road above and was restored by contractors but with considerable co-operation from the Inland Waterways Association and the two boat clubs on the canal. The gauge marks the new internal profile of the tunnel, 1989.

This breach at Llanfoist effectively bifurcated the Brecon & Abergavenny in 1975, the canal being closed for through navigation for six years. The extent of the collapse is all too visible here. *British Waterways Board*

The draw bridges on the Brecon & Abergavenny canal range in use from solid road bridges to lighter farm tracks. *(Top)* near Pencelli in 1989; *(Below)* is at Talybont-on-Usk in the 1970s. *(Bottom) L.A. Edwards*

Chapter Five

Future Infinitive

Of the Aberdare Canal very few relics are to be found, as most of the waterway was obliterated by road-building. The sole relic that was still extant in 1974 was a group of cottages at Mountain Ash (ST 048992), these now backing onto the road. The last traces of the basin at Abercynon (ST 085948) where it made a junction with the Glamorganshire have finally disappeared under a new fire station. It was fortunate that after such a frustrating search the 'Navigation', still extant, was still plying motorists with ale in lieu of the boatmen who were once welcome there.

Save in detail, the Brecon & Abergavenny is virtually complete, barring only a short length in Brecon itself. The terminal warehouses still stand (SO 044284), albeit now facing a car park. A far better spot to start is by the gasworks (SO 048282). At a nearby cottage is a milestone beautifully preserved and repainted. Alongside the gasworks wall enthusiastic hunters will find mouldering away the old 'M.R & C.C^{o}' boundary markers:those in Newport antedate these with 'M.C.C^{o}' lettering.

To see this canal's bridges a good location is at Glanusk Park (SO 196190), although during draglining operations carelessness led to the breaking off of the lovely mileposts. Fortunately, and surprisingly, the old ironbridge number plates are still *in situ*.

The prettiest lock is undoubtedly at Cwm Crawnon (SO 144198); while the aqueduct at Gilwen (SO 244144) over the River Clydach has easy access and great charm.

For the rest may we suggest a cruise? This is the only canal in South Wales where you can cruise throughout and it would help to prove to the owners, the British Waterways Board, that it is worth keeping.

There is little doubt that some Welsh (or English) authority is determined to obliterate the relics of industrial Wales and specifically the Glamorganshire Canal. Success is apparent, for of what was once a 24½ mile canal, little remains. Perhaps it is this that makes the whole exploration so unpleasant, although in the Merthyr Vale the Welsh themselves seem depressed, a reflection on the tragedy of Aberfan through which village the canal once passed.

Cyfarthfa Castle, once the home of the Crawshays, is well lit (365 windows) and contains an interesting museum, albeit of the 'do not touch' variety. Once the canal ran near the castle, but until Abercanaid (ST 060038) nothing of the waterway remains. Here, aside from two docks or basins, a fixed swing bridge – with interesting expansion brackets – and a lockhouse, remote in the fields, there are still two weirs and 'Quay Row', a batch of cottages. Nearby too are a couple of old cannon doing duty as boundary markers. At Abercynon (ST 086950) the next nodal point, little remains save the ruins of lock chambers, while at Tongwynlais (ST 132822) although the canal is ostensibly a part of a nature reserve, it is dismal and, at least in late 1973, somewhat polluted. The waterwheel at Blakemore's Melingriffith Works (ST 144803) is still extant although it appears to be

The Brecon & Merthyr railway bridge and road bridge over the canal at the rear of the White Hart Inn at Talybont, looking south towards the Talybont reservoirs.
I. Williams

The canal wharf at Brecon with the tramroad arch bridge of the left. *D.S.M. Barrie*

desireable to obtain permission to view. The canal thereabouts is (untypically) attractive in a pastoral manner.

On 27th January, 1956 the 'Special Correspondent' of the *Cardiff and South Wales Times*, under the heading 'Child Killer of Tiger Bay' sought to fill in the canal as it was a danger to children, with dramatic phraseology: 'Every time I see the lad who by the mercy of Providence I managed to save, I wonder how long it will be before it is made possible for that canal, black, brooding, sinister, to claim another victim'. Perhaps it is justice that the best part of the visible remains of the Glamorganshire Canal, the weighing dock, is no longer at Cardiff, but instead at the Waterways Museum, Stoke Bruerne, Northamptonshire, England (SP 743500).

Of the Monmouthshire Canal in Newport, singularly little remains, and it is best to pick up the line at the M 4 overbridge (ST 305895). However beware of parking nearby; while the sight of people piling out of the chapel into the pub is amusing, not so are some of the rougher, pugilistic and vandalistic natives. Under the motorway, notwithstanding the 10 ft headroom, the canal is sealed by a row of steel piles and a walk up the main line is a somewhat chastening experience. The quiet pastoral charm of a disused canal contrasts poorly with the frenetic rush and roar on the motorway. However the bridges, locks and milestones are well worth the walk right through to Cwmbran (ST 293949). The stretch from here to Pontypool with 'cascades' instead of locks, bedsteads in lieu of boats and disinterest for enthusiasm, is dreary in the extreme. Nevertheless there is yet hope that from Five Locks Road to Pontymoile (SO 294004), the latter a charming spot, restoration may be possible. The primary obstruction is at Crown Bridge, Sebastopol, where a culvert prohibits boating, a nastiness that was fought bitterly by the Inland Waterways Association, South Western Branch, under its doughty Chairman, Fred Blampied; but they shall win one day and we may yet see boats through Cwmbran Tunnel once more.[25]

If you seek beauty it may be sensible instead to follow the Crumlin Branch from the M 4, through Allt-y-ryn to the foot of the Rogerstone locks (ST 285885). *En route* a charming wooden Great Western Railway notice reminds us that 'BATHING in this canal is STRICTLY FORBIDDEN. Any person found bathing without permission WILL BE PROSECUTED', but innumerable children nearby obviously could not read.

The Monmouthshire Canal, although connected to the Brecon & Abergavenny Canal, has not been so lucky with its restoration schemes. When re-opening was first propounded the waterway was in a very poor condition and it was felt that not only would restoration be hopelessly unviable but for various engineering reasons, impracticable. Underlying the whole story is the unspoken feeling in that area of South Wales that the pleasure craft had the Brecon & Abergavenny and that the fishermen wanted a canal all to themselves undisturbed by boats. Fortunately perhaps, a compromise has been reached whereby on separate stretches unpowered craft will be permitted, picnic sites made or fish-ponds stocked. Behind this compromise was a struggle in which Mr M.J. Bevan, Secretary of the Risca, Magor & St Melons Canal Preservation Society, was to play a leading part.

A mixed freight train crossing the unique structure of the Crumlin Viaduct as viewed from the towpath of the Monmouthshire Canal.
British Rail

The stage was set when at a public meeting held on 9th May, 1968, it was unanimously agreed to make a concerted defence of the waterway through the area in the face of the proposed new Risca by-pass which would have entirely obliterated most of the canal. The inaugural meeting of this society took place on Thursday 30th May, 1968, when about 30 people attended and decided that voluntary working parties should be provided to assist Mr Lloyd, the British Waterways Board's Area Engineer, in maintaining the canal; that all reasonable publicity should be sought to further their ends and that they should be affiliated to the Inland Waterways Association. Particular assistance was received from the Pontymister Angling Club who provided a most useful graph of the canal, giving the minimum and maximum depth of water between Pontywaun (ST 220926) and the Giles aqueduct. This piece of foresight on the part of the Angling Club could well be copied by many others throughout England and Wales, for too often no-one is aware of an obstruction shoal until a boat is fixed firmly on it.

The society, as evidenced by the minutes of their meeting held on 4th July, took bold and imaginative steps to gain publicity, taking a cine-film of the canal – from a boat! and slides for use at lectures; and with a float at the Risca carnival they won first prize. At this date it was agreed that approaches be made to the local and county councils to see whether, under the Amenities and Recreation Act, they could take over the waterway.

Subscriptions were to be kept to 12½p p.a., relying on donations for working capital. At the next meeting, in September, it having been ascertained that an insurance premium of £11 10s. (£11.50) was necessary to cover any 30 persons in a working party on the canal at any one time, it was agreed to obtain coverage in order that work might be carried out from Pontywaun bridge to Green Meadow bridge, adequate tools having been already obtained. Some weeks later an attempt was made to lease 1/2 Halls Road Cottages near Pontywaun bridge, which were owned by British Rail who subsequently replied that these were still under negotiation with another body. In 1973 these were still under negotiation albeit they were now vandalised to dereliction. The Forestry Commission on the other hand agreed that any silt dredged from the canal could be deposited on their land, a great advantage as dredging waste disposal is one of the greatest problems facing any canal management.

By the 4th of December a punt, latterly the property of the British Waterways Board, was being repaired to assist working parties and the old canal maintenance shed at Pontywaun was available for their use.

The Giles aqueduct, which had deteriorated badly over the years, was by January 1969 in some danger of being completely eliminated. Again restoration was impracticable, but had the aqueduct been completely severed there is no doubt that there would be an insufficient supply of water to maintain even the St Mellons section and the society felt obliged to press with all their power for the provision of a by-pass pipe.

Another matter running as a theme through the Society's minutes, was the condition of the canal towpath. The British Waterways Board (BWB) accepted that they were obliged to maintain the towpath in good order despite the mess caused by their workmen engaged in leak-stopping, obstruction by coal lorries or its use as a rubbish tip.

The first voluntary working party took place on 1st March, 1969, and 25 workers turned up 'to clean the canal at the right of the Green Meadow Bridge'. At about the same time four park benches were installed on the towpath each bearing a plaque indicating that they were donated by the canal society. Regrettably, although theoretically vandal-proof, within a very short time this proved not to be so and now only the stumps remain. A little known requirement was that the BWB had to have an indemnity releasing them from 'any damage incurred in respect of the seats being placed on canal property' and, in fact, the society then found it necessary to take appropriate insurance coverage against any risk.

At a meeting held on 25th June, one most cheerful note was that the Clerk to the Magor & St Mellons RDC had advised them that 'the Society would receive a donation of £150'. Also on 25th June it was requested that the Risca UDC supply the society with six pairs of waders, which was agreed. On 6th August, as rubbish tipping was becoming progressively worse, the society approached the Council to put a notice in the local papers stating 'that it was illegal to tip rubbish [in the canal] and that offenders were liable to prosecution', which the Council confirmed they would do.

At the beginning of January 1970 the Risca UDC's Surveyor stated that the Council would take over the stretch of the canal within their boundaries 'within the next month or so'. On the other hand the Usk River Authority had stated that they would charge £4 per million gallons of water (approximately £1,500 p.a.) that flowed into the canal. On 15th–16th December dredging work had been undertaken at Adams Quarry and although the machinery was unsuitable it was stated that the operator did well, furthermore the Society agreed to assist the BWB financially in any dredging work they undertook, £50 for the Risca section and £50 for the St Mellons. It was also suggested that the BWB might care to contribute to the purchase of nine tons of shale for towpath work.

Dredging work, paid for by a member of the Society, was undertaken at Risca Quarry from 16th to 21st March, achieving a depth of 5 ft in the middle of the canal tapering off to 18 in. on the towpath side and 6 in. on the off-side.

At a meeting held on 23rd March, the Risca UDC stated that they proposed to spend £20,000 in one year on the restoration of the canal, with the British Waterways Board offering them half of this amount if they accepted ownership. It was also made clear at a meeting on 8th April 'that the Society was requested to refrain from undertaking major redevelopment schemes while the local authorities negotiated with the British Waterways Board'. Dredging work had been undertaken above Rogerstone Locks (ST 27886) which involved the Society in an expenditure of £420, £90 of which would be reclaimed from the Magor & St Mellons RDC. A curiously negative note also appeared at this meeting when it was agreed that 'all riding schools in the area of the Fourteen Locks be written to asking for their co-operation in refraining from using the canal towpath.'

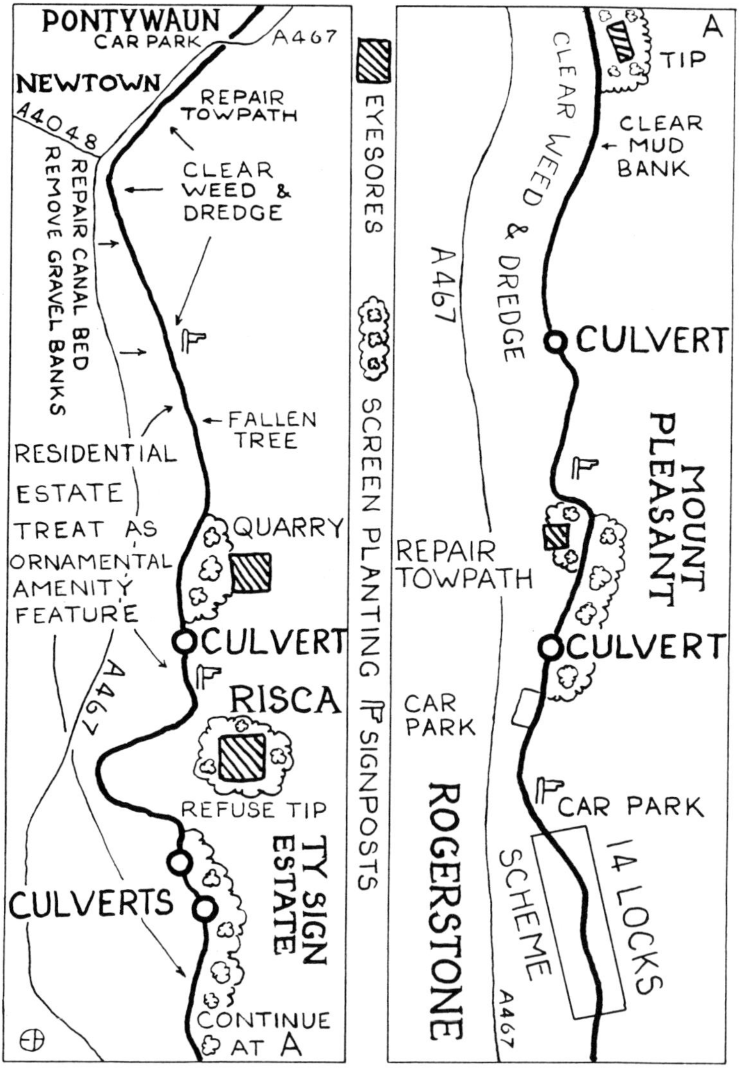

A diagrammatic sketch of the Monmouthshire Canal amenity dream of the early 1970s. Some work has been done but it is unlikely that this plan (*see also page 98*) can ever be completed.

Another underlying current within the meetings was that of difficulties with Laings who were building the Mount Pleasant Estate (ST 262896). Thus we find on 26th May, 1971, 'Laings had deposited a considerable amount of rubbish in the canal at the estate', a matter which was taken up with the local Council. The Risca Quarry reappears on 9th July; concern was expressed about the industrial waste filtering into the canal from these works and when it had been explained that the quarry had 'existing use rights' it was agreed that the stagnation and unhealthy odour should be referred to the public health inspector. At the same time there appears a note on a typical sort of struggle between conservationists and the public, as it was proposed to request the BWB to instal metal stakes along the towpath at the Rogerstone flight 'to prevent its use as a vehicle lane'. Similarly at a meeting held on 28th January, 1972, a member said that infilling the canal at Fernlea [which has subsequently taken place] would not alleviate road traffic problems but that it would help the flow of water if the Fernlea feeder could be diverted totally into the rest of the canal.

By 29th October, 1972, various stretches of the canal around Pontywaun, Mount Pleasant and Adams Quarry had been cleaned out by the Risca UDC, but in the same month another of the problems any preservation society can face was raised by residents of the Border Homes Estate – they were worried about rats. The society agreed to write explaining the difference between rats and voles and pointing out that this problem was really one for the council to deal with, although a news item in the *Western Mail* as late as 27th September, 1973 proves all is not well: 'People living in Mount Pleasant Terrace, Pontywaun, near Cross Keys, claim 15-inch rats are breeding near the Monmouthshire–Brecon Canal, which has recently been cleaned out'.

By July 1972 the work of the Society as a watchdog-body was, to a certain extent, completed when the Newport Borough Council accepted a report on the works to be undertaken within this area. Effectively this council would become freehold owners of 3.5 miles of waterway involving 9 locks and 8 bridges, but split into three segments. Their objectives, simply, were:

1. Safety. To minimise or eliminate wherever possible, drowning and falling hazards.
2. Economy: To produce the most economic possible scheme for the restoration, consistent with safety considerations and the 'maximization' of amenity and recreational opportunities.
3. Amenity: To maintain wherever possible the essential 19th century character of the canal and its rural setting.
4. Water Surface: To utilise the recreational use of the limited water surface consistent with the preservation of the historic and rural character.
5. Footpaths: To maintain existing footpaths and to develop new links wherever possible so as to increase the accessibility of the canal to local residents.
6. Car Parking: To provide adequate off-street parking facilities and to minimise danger and disruption to main road traffic.
7. To reconcile conflicting demands and interest of recreational users.

The Council faced many difficulties, as not all activities are compatible; for example, horses, bicycles and pedestrians cannot co-exist on a towpath, and most assuredly boating and fishing interests cannot co-exist. Unfor-

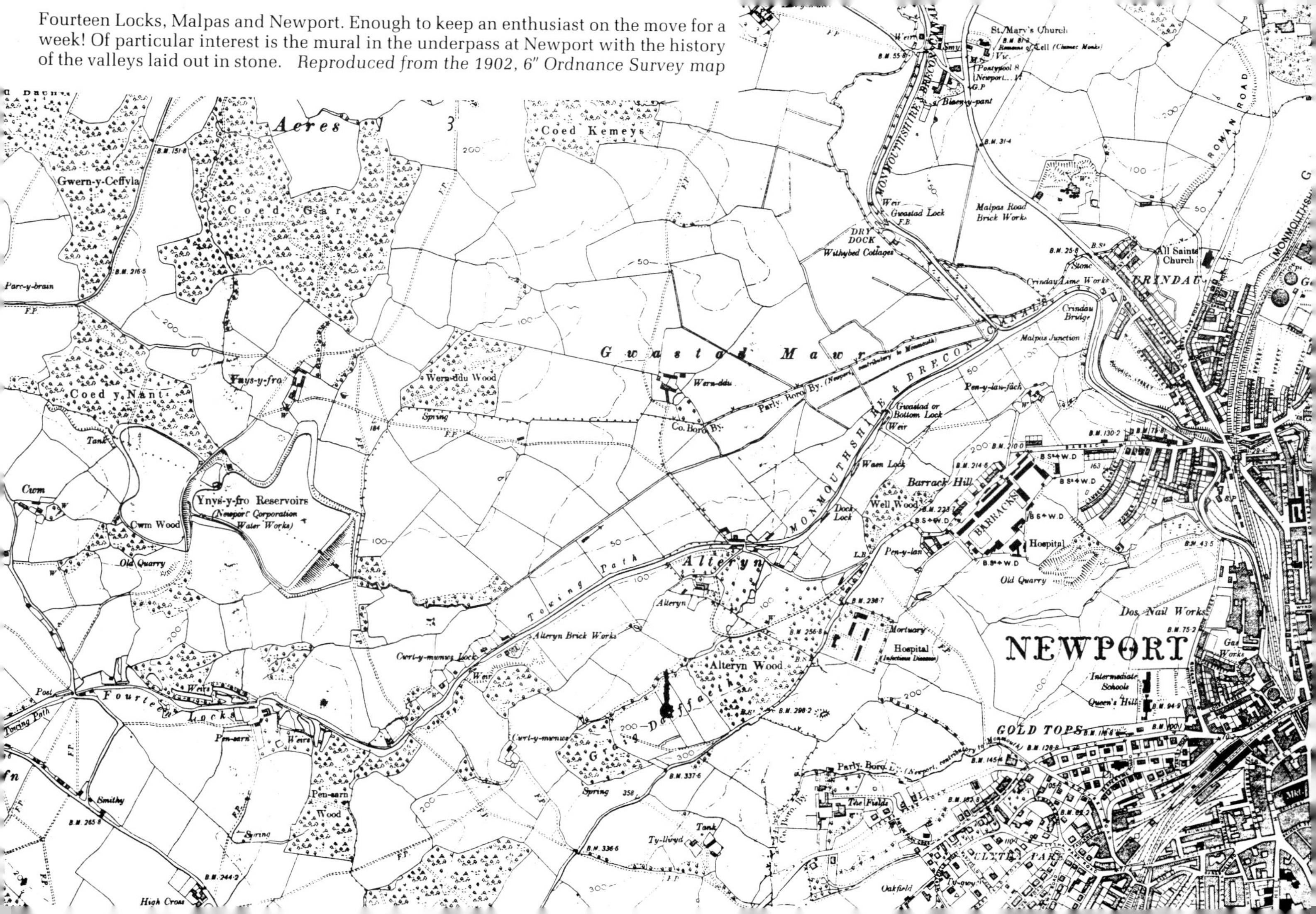

Fourteen Locks, Malpas and Newport. Enough to keep an enthusiast on the move for a week! Of particular interest is the mural in the underpass at Newport with the history of the valleys laid out in stone. *Reproduced from the 1902, 6" Ordnance Survey map*

Locks on the Monmouthshire canal measured 64 ft 9 in. × 9 ft 2 in., the canal channel being dredged to a nominal 3 ft depth. At Rogerstone the 14 lock Scheme had been, for over 15 years, the subject of much planning, considerable debate and apparently acrimony between the different bodies who are or were involved in its beautification. These two views show the general state of the lock chambers in 1989.

tunately, the most practical part for boating is the northern arm running from Malpas Junction to the borough boundary but this is also the best fishing water. Again, nature conservation interests oppose removal of weed whereas both anglers and boaters cannot abide it. On one matter the council had a definite opinion – no powered craft will be permitted on the water in Newport. Lock treatment was to be taken in stages. Cascades, a series of steps over the length of the lock providing a waterfall effect which, being estimated in 1972 as costing £2,066 each, would be used on the Crumlin Arm from Malpas Junction to Cwrt-y-Mwnws. Boat rollers, effectively an inclined ramp with 4 ft wide rollers in the middle and steps on either side (£4,200 each) would be installed at Ty-Ffynon Lock as it was considered that this lock was too remote to manage adequately and supervise lock-operation. Curiously, the restoration of locks, new gates, repairing and cleansing of sluices, pointing and general repairs would have cost the identical sum of £4,200 each, but an additional capital expenditure of £3,240 per lock was thought to be necessary to pump water from the lower to upper ponds, this sum allowing for 20 boat movements per day. Maintenance costs would, naturally, be higher, but it was recommended that Gwasted lock should be so restored while a decision on Malpas Lock was postponed for the nonce.

In January 1973 at the other end of the canal, the Risca section, Pontywaun to Welsh Oak, was being cleared of waste and weed growth and the Giles Aqueduct was being demolished. On the Magor & St Melons section, Welsh Oak to Rogerstone, the sum of £23,170 to be spent by 30th June, 1973 had been allocated by the Welsh Office under the Environment Scheme, for dredging and other purposes. It was originally proposed that the water level within the old lock pounds be held at 18 inches so as virtually to eliminate any risk of drowning but upon protests from anglers new proposals did allow for 3 ft of water.

The 14 locks at Rogerstone (ST 284885) are to receive a totally different treatment from the Newport end. Again the conflict reappears between retaining the industrial character of the flight of locks and providing facilities for informal outdoor recreation. It was stated in the report on the locks that 'these are deep, 17 ft, and are danger points acting as magnets to children'. The alternatives offered are:

1. Shallow water in locks with dams at the bottom and providing waterfalls. Said to be visually exciting but not dangerous if proper guardrails are provided.
2. Culvert the lock and infill, it being then 'very safe indeed' [*sic*].
3. Stepped cascades.

The backbone of the whole 'Fourteen Locks Scheme' is a route for pedestrians where it was envisaged that they will walk through a series of outdoor 'rooms', with the canal locks and their pounds acting as visual and physical links between. The 'rooms' will be quiet areas, usually natural pausing or resting places along the 'corridor'. Shades of the old boatmen's and leggers' huts at the end of canal tunnels!

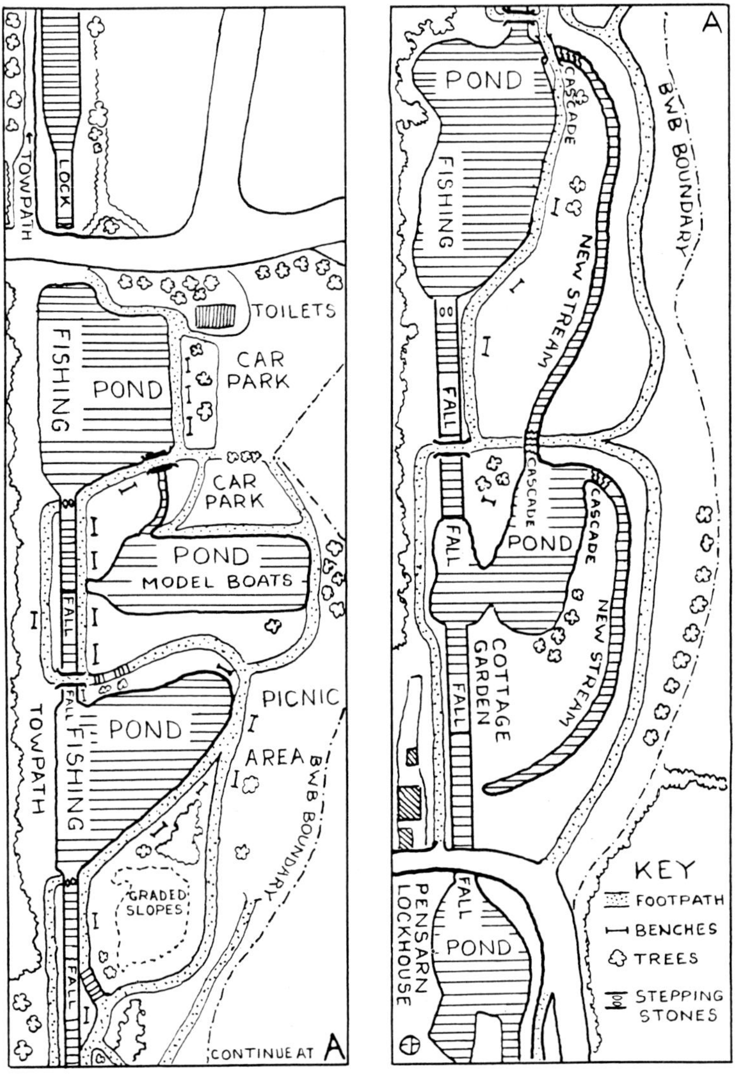

Diagrammatic sketch showing part of the 1970s Monmouthshire Canal Fourteen Locks scheme. Never completed, but it was a great plan (*see also page 93*).

Many car parks are to be provided and it is pleasant to note that provision is to be made for the elderly or disabled to get near to the canal without leaving their vehicles. No overnight parking, heavy vehicles or mobile shops will be permitted. Litter-bins and their collection was considered to be a matter of importance and a Warden would be appointed to cover the whole area from Pontywaun to Newport, although Welsh vandals being in no way different from any others, one suspects stronger measures will be required if the whole site is not to be reduced to wreckage within a very short time.

One curious juxtaposition, mainly for economy, is that a display-stand giving information on angling, bird-watching, canal history, the working of the locks and the boats, is to be built onto the toilet block.

The Neath Canal now serves as a water channel to feed the works in Neath, especially the British Petroleum refinery. To ensure that there are no difficulties the entire shareholding in the Neath Canal Navigation is held between the factories who have interest in the continuance of water supplies. So vital is this waterway to the industry of this valley that it was said a single German bomb on Ynysbwllog aqueduct would have stopped all work. As the water must be kept as far as possible unpolluted, no craft are allowed on the navigable stretches, the locks being either weired or pinned open. Two of them have most ingeniously been converted into boathouses to give some protection against vandalism for the maintenance craft. Rubbish tipping and weed growth are the Manager's two main problems, the former resulting in an unfortunate, but unavoidable, policy of letting the towpaths become almost impassable.

Physically, thanks to the work of the foremen including Will Evans (75 and going strong) and their gangs – all ex-colliers and, therefore, accustomed to work – the water channel is in excellent condition from Resolven through to Neath, as is that of Giants Grave Canal. The sector Glyn Neath to Resolven is completely overgrown and due to be converted to a road, despite its charm and interest. Points of interest on the line of the canal include the aqueduct (SN 803011), the workshops restored recently, the terminus in Neath (SS 734941) overshadowed by the gaunt viaduct of the A483.

The Smith (or Llansamlet) Canal is now little more than a wet, dissected ditch. However there are still bits and pieces to be seen on the ground, including the odd bridge, scattered alongside various factories and railway lines on the outskirts of Swansea. One portion is visible near the Addis Factory (SS 667953) and another at Church Road, Llansamlet (SS 685976) and of considerable interest is the Hafod Copper Works nearby where there are walls built from moulded copper slags – an unusual use for a waste product.

The Swansea Canal is, alas, doomed to extinction. Of the terminal basin at Abercrave (SN 820125) singularly little remains, being bulldozed flat in 1973, but by following the A4067 Sennybridge–Swansea road to Ystalyfera (SN 772090) behind what was the cinema may be found a delightful, albeit exceedingly porous aqueduct, over the Afon Twrch. Only odd traces still occur from there to Ynysmeudwy, but here is the Swansea Canal *par excellence*, immediately by, and accessible from the A 4067 are the relics of the

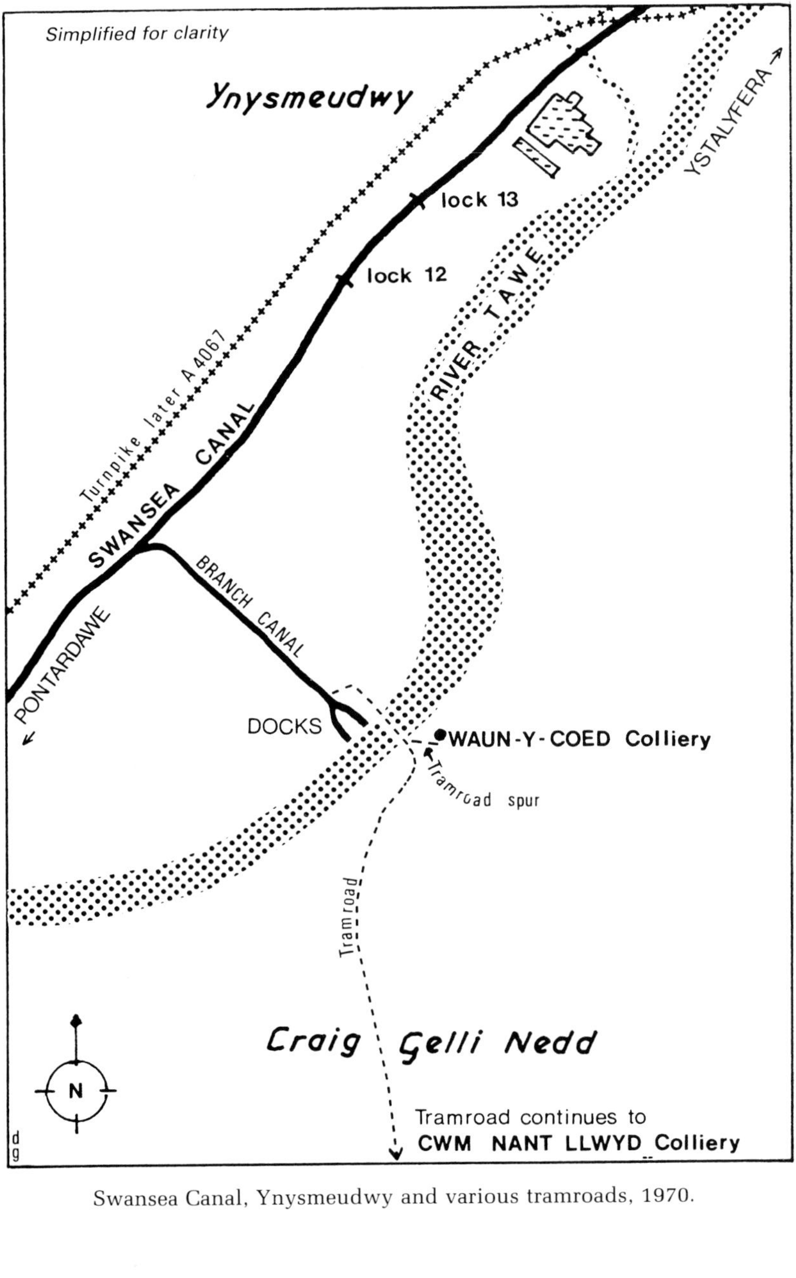

Swansea Canal, Ynysmeudwy and various tramroads, 1970.

pottery, latterly a tinplate works and within a quarter of a mile the chambers of locks 13 and 12 with, tragically, a lonely rack poking up from a hole in the ground. A short walk along the towpath and the old branch canal is to your left, which at its double basin made a connection with the tramroad to Cwm Nant Llwyd Colliery. Interestingly, a length of Brunel bridge rail still lies in the dried up basin. Incidentally the works, which at various times saw the manufacture of terra cotta work, bricks, exquisite china and only latterly tinplate products, would be well worth exploring before it is demolished; until mid-1973 there was an old crane, a heap of bridge rails and of tram-plates; all have now disappeared virtually unrecorded.

From this point if one wants the essence of the Swansea Canal it is best to follow the towpath through to Clydach, but for motorists the next stop is at Clydach (SN 687008) where, behind the Gentlemen's convenience, immediately adjacent to the A 4067, is a single-arched aqueduct, somewhat in need of repair. To the left is a church, mill and warehouses, to the right the back scenes of Clydach. Crossing over to the other towpath and following the canal in the Swansea direction one is led to a glorious foundry complex; in the yard, somewhat incongruously, lies mouldering a Wilkinson-type enclosed tram engine. Once a major railway transhipment point, the canal is now sullen and somewhat pungent, while below the bridge it is filled with rubbish and almost finally gone.

From here to Swansea there are, or were, lengths of canal; but the advance of all-obliterating 'progress' ensures that any notes would be misleading.

Down the road from the Neath Canal, the Tennant also serves as a water channel feeding factories in Neath and Jersey Marine; although remaining in private ownership. Somewhat less accessible than the Neath, points of interest are a superb aqueduct near Aberdulais (SS 773993), complete with its regulation lock. Although somewhat lost beside the railway viaduct, the engineering of William Kirkhouse remains as sturdy as ever. Some interest may also be found in the stretch of waterway across the Crumlin bog (SS 700936), if only as a rare example, in the British Isles, of such canal building. The disconnected terminus at Port Tennant, although changing, has some unusual buildings at SS 675928.

The main problem faced by Welsh Valley canals was their very success. The shareholders were, rather than individuals, the great ironfounders, copper refiners and coal-owners, the former wanting their raw material available at the works as inexpensively as possible and all parties, in a fiercely competitive market, needing the finished products moved to the docks quickly and economically. It is sad, but true, that lives were cheap – and few cheaper than on the tramroads or even the waterways – and there seems always to have been a conspiracy to hide this price, the turnover of Irish navvies being higher even than in England.

Within a year or two of their opening the canals, almost without exception were hopelessly congested with 24-hour working the rule rather than the exception, although thankfully family boats, as used on English canals, seem not to have officially existed. It is not coincidental that the owners of the waterways were among the forefront of those anxious to build steam

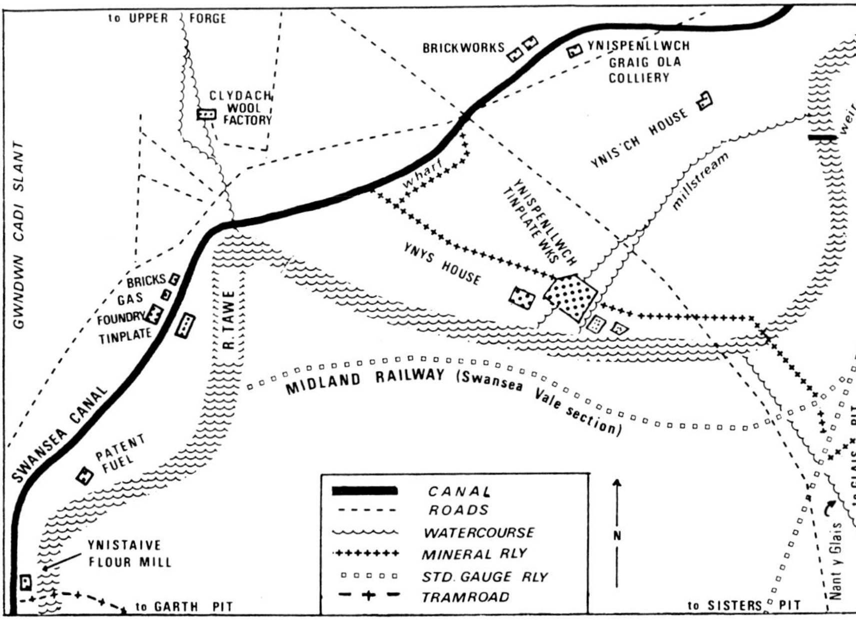

CHANGING FACE OF CAPITALISM CLYDACH; above 1875; under 1950

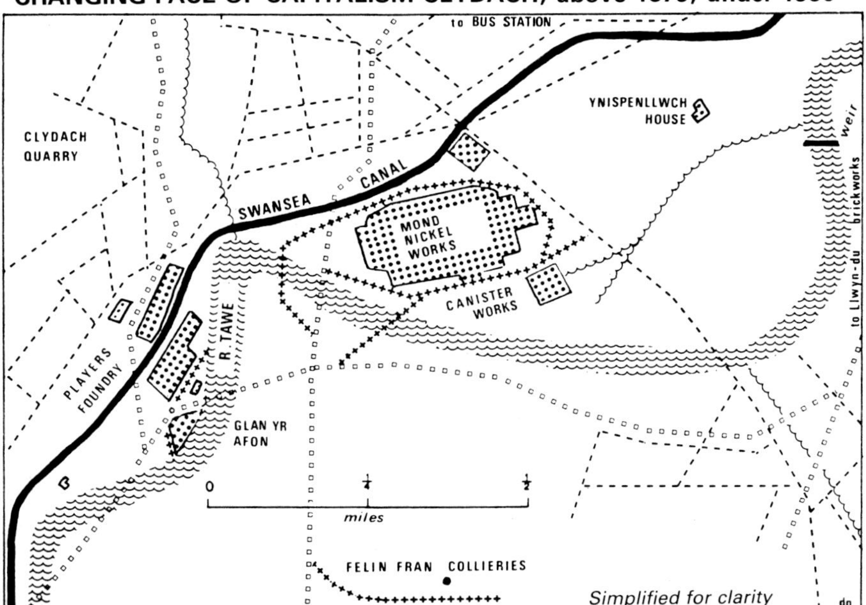

railways parallel to, or along the bed of, their canals. In 1842, and of the Monmouthshire, it was said 'The Inconvenience and expense to the Freighters of the present defective means of Communication by canals and tram-roads, are so well known, that no vindication or apology for the present undertaking [for a steam railway] is required from its promoters', and within a year the Monmouthshire Canal Committee noted that the proposed railway's Chairman believed 'That a railroad between Newport and the Pontypool Iron district is desirable, as even some of your Committee do, I know, admit'.

Welsh canals, albeit still functional even in the 1920s, have been long overshadowed by railways and, with space in the Valleys at a premium, were fair prey to any rail or road scheme, so their survival is rather miraculous. And in very many ways the canals and tramroads of the Welsh Valleys differed from their English equivalents. There were none of the catastrophic failures that bankrupted ordinary members of the public, as was the case with the Dorset & Somerset, Grand Western or Kingston, Leominster & Stourport canals which were planned to run through not dissimilar countryside. The contractors and engineers were those who before, after or even during their Welsh activities worked on English waterways, using exactly the same engineering techniques, with some successes and not a few minor failures. If one compares the Welsh valley canal aqueducts with, say, Pontcysyllte or even Wootton Wawen there is no comparable inspiration, but by the standards of Dundas or Great Hayward, Aberdulais is a state of the art structure.

Locks were, generally, of an odd gauge, boat building methods anachronistic and, in the early days at least, cargo handling methods crude, there being little incentive after the 1840s to alter them. Lock gearing is an example, for single tooth rack and pinion sets were replaced by 'double draw' in the early 1800s, but survived in Wales until today.

And this difference has spilled over into the modern use of canals in the valleys. For example the restoration of the Kennet & Avon in England has reached its culmination and by any standards it has been a magnificent effort; the Fourteen Locks Scheme at Rogerstone has been running for nearly as long but goes two steps forward and one back. The Kennet & Avon has its own Association, the Fifteen Locks has been looked after by a number of canals societies. Similarly, restoration of the Pocklington Canal (9½ m/15 km), again looked after by one body, the Pocklington Canal Amenity Society, has altered it from a disused, and hence dangerous, muddy ditch to a bright, wholesome, amenity, open to all. The Neath (13 m/21 km) has hitherto suffered from the seemingly irreconcilable requirement of its owners, the great petrochemical industry, for a water supply – nothing more or less – and the pressure of those who want the channel for boating and other amenity purposes. For many years the late W. Porter, the Manager, aided by his foreman, the late Will Evans, and a small gang of men, tried desperately to keep the channel open and at least restorable, often paying for some works from his own pocket. Now the Neath & Tennant complex is effectively bifurcated with the length from Resolven to Ysgwrfa owned and being restored (excellently) by the Neath Canal Joint Management Committee – the Borough and County Councils plus the Neath & Tennant Preservation Society united in one. The bottom half is merely ticking over.

While it may be said that there is more money available for the Kennet & Avon than for the Monmouthshire & Brecon, to use these canals' modern artificial title, the same cannot be said of the Pocklington *vis-á-vis* the Neath; there is more money in the Valleys now. Therefore we can only summarize developments in the Valleys since the first edition of this book as follows.

The Neath & Tennant canals, now regarded as one, will undoubtedly, one day, be restored to a great extent, but for now the Joint Committee are operating under two licences. The Neath Borough Council have obtained a 10 year licence (from 1986) from the Neath Canal Navigation Company for the 'purpose of cleaning, repairing, maintenance, and using the canal for leisure purposes' between Ty Banc Cottage (400 m above Resolven Lock) and Yscwrfa Bridge and including the towpath from Resolven Lock) and Yscwrfa Bridge. The Council has a similar agreement with the Welsh Office for the canal cutting owned by them, between Resolven Bridge and Ty Banc Cottage. The plans which will undoubtedly affect the programme further along the canal include the use of redundant railway bridges as part of circular walking tours, with the towpath restored to full use, fishing (with facilities for the disabled) and nature conservation. Industrial archaeology is said, rightly, to be 'highly desirable in terms of developing the canal's image' and, pleasantly, the provision of picnic sites enables parties to relax and enjoy the pleasures of a waterway. Boating, at present, is subject to a number of restrictions, and reasonably so for waterborne pollution can quickly undo many months work. However a Rainbow boat, designed to carry disabled parties (mainly from special schools) entered service in mid-1989.

The Swansea canal represents a hard challenge for its would-be restorers, but at present work is concentrated on Ynysmeudwy Basin at Pontadawe and its adjacent tramways, this area being one of the most attractive and interesting to be found on the valley canals. Interestingly, the length from Ystalyfera to Clydach was declared a Remainder Waterway and this length fulfils such a water supply requirement that it pays for its existence and labour force. Plans exist for the restoration of as much of the canal and its tramroads as possible, but the Swansea City Council and the canal society alike are sensible enough not to take on too much work at once.

Viewing restoration of the two canals, the Brecon & Abergavenny and the Monmouthshire is to see an incredible contrast. From North of Bridge 66 just short of the original terminus, for roughly 33 miles, to the Crown Inn adjacent to Bridge 48, we have an example of a beautiful canal abounding in interest and with a fair degree of tranquility. There are problems insofar as the hire boat companies are uniformly worried, there being an excess of boats to the number of would-be hirers. Moored boats, too, are appearing in excessive numbers and it has been suggested the continual hammering of boat mooring spikes into the banks will, eventually, seriously weaken them. Nonetheless that portion of the Monmouthshire and Brecon has a foreseeable future.

Only a mile or so away from the Crown is the site of Pontnewydd Locks, some erased and others turned into a water garden. Bridges are lowered and the canal piped, although a half-mile immediately above the locks is maintained as a nature resort and linear dump for the local populace. In passing, one must mention Cwmbran as the only place where we were physically stoned by youths while photographing the canal.

Plans are in existence to re-open both the Pontymoile to Newport and Crumlin lengths of the canals, but, by contrast the Torfaen Museum Trust's display at Junction Cottage, Pontypool, has been closed for a couple of years following a dispute with the British Waterways Board over payment for essential repairs. In 1986 the souvenir brochure of the Welsh Waterways Festival Rally, held at Pontymoile, stated 'The Monmouthshire, Brecon & Abergavenny Canals Trust is now carrying on the good work started by bodies like the Torfaen Canal Society and the Newport Canal Preservation Society. The borough council figures strongly in the development strategies and plans now being put into effect, and the future of our Canal seems assured'. But the famous Fourteen Locks Scheme really has little to show since it was opened to the public in 1976, the whole site, Museum and displays, calling for extra funding if it is to match the dreams of 1972, and unless this funding is forthcoming soon the continued erosion of the waterway track will negate any attempts at restoration. One also wonders if there is the will to expend the time, money and enthusiasm necessary for any revitalizing of a waterway; the speed of the works at Neath shame those at Newport.

Appendix One

The Neath Canal

The 'vade mecum' of historical waterways – *Bradshaw's Canals and Navigable Rivers* (1904) – has one particular omission relating to Welsh canals: the lock and bridge names of the Neath Canal. As many of these are doomed to disappear soon their names are given below.

Reading from Neath upwards

Locks		*Bridges*	*Bridges (contd)*
No. 1	Lock House	Iron Works Bridge	Aqueduct Bridge
2	Lock Machine	Ship Bridge	Ynysbwllo Bridge
3	Witworth Lock	Pont David Bridge	Ty'r Bont Bridge
4	Lock Gitto	Swansea Bay Bridge	Ynysarwed Bridge
5	Ynysarwed Lock	Tricks Bridge	Abergarwed Bridge
6	Abergarwed Lock	Metal Box Bridge	Farmer's Bridge
7	Farmer's Lock	Galve Bridge	Resolven Bridge
8	Resolven Lock	The Green Bridge	Crugiau Bridge
9	Crugiau Lock	Bridge Street Bridge	Rheola Bridge
10	Rheola Lock	Ash Tip Bridge	Aber-clwyd Bridge
11	Aber-clwyd Lock	Lock House Bridge	Maes-gwyn Bridge
12	Ynisultor Lock	Aberdulais Bridge	Llena Bridge
13	Maes-gwyn Lock	Taylors Bridge	Bwllfa'r Onn Bridge
14	Bwllfa's Onn Lock	Pont-y-gwaith Bridge	Aber-pergwm Bridge
15	Granery Lock	Ynys-nedd Bridge	Screens Bridge
16	Chain Lock	Witworth Bridge	Foxes Bridge
17	Foxes Lock	Clyne Bridge	Lamb & Flag Bridge
18	Lamb Lock		Maesmarchog Bridge
19	Maesmarchog Lock		

Appendix Two

The Fourteen Locks Scheme

Monmouthshire Canal – Fourteen Locks Scheme

Angling Data: 4 lock ponds to be set aside, 2 ft 6 in. deep at the edges and not exceeding 10 ft in the middle. These are:

	Recommended stock		
Pond No. 1 (3/4 acre, rectangular)	Roach	30 lb. =	60 @ 1/2 lb.
	Perch	30 lb. =	60 @ 1/2 lb.
	Carp	50 lb. =	4 @ 5 lb.
			15 @ 2 lb.
	Tench	35 lb.	35 @ 1 lb.
	Bream	35 lb.	35 @ 1 lb.
Pond No. 2 (3/8 acre, triangular)	Roach	15 lb. =	30 @ 1/2 lb.
	Perch	15 lb. =	30 @ 1/2 lb.
	Carp	26 lb. =	2 @ 5 lb.
			8 @ 2 lb.
	Tench	17 lb. =	17 @ 1 lb.
	Bream	18 lb. =	18 @ 1 lb.
Pond No. 3 (1/3 acre, lens shape)	Roach	30 lb. =	60 @ 1/2 lb.
	Carp	50 lb. =	4 @ 5 lb.
			15 @ 2 lb.
Pond No. 4 (1/4 acre, round)	Tench	30 lb. =	30 @ 1 lb.
	Bream	30 lb. =	30 @ 1 lb.

Total 180 lb. Estimated total cost (1972): Fish £500
Ancillary Works £200

'Some weed on, and in, the ponds would be necessary as cover from predators, i.e. anglers [*sic*], herons, otters.'!

Monmouthshire Canal, Rogerstone '14 locks', two-rise staircase.

Pontnewydd or Cwmbran locks, Nos. 57 and 58, have been landscaped so as to preclude any likelihood of navigation. Now an ornamental feature of one of the town's parks, the local youths find cans placed on the steps useful for air-gun practice.

Appendix Three

Newport Gas Works

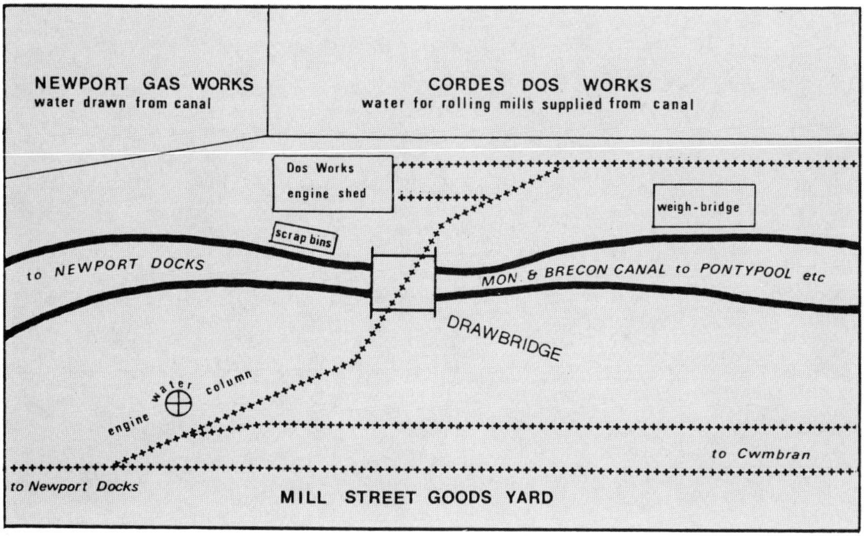

The movable and unique skew drawbridge across the Monmouthshire Canal at Mill Street, Newport as seen in GWR days.

Appendix Four

Statistics

Iron carried on the Glamorganshire Canal

Name of Works	*1817–9*	*1820–2*	*1823–5*	*1826–8*	*1829–31*	*1832–4*	*1835–7*	*1838–40*	*Totals*
Guest & Co. Dowlais	30426	38243	42470	60902	73074	97944	117345	125074	585478
Crawshay Cyfarthfa & Hirwain	46543	54217	62914	79529	60125	97000	103324	109502	613154
Thompson Penydarren	24658	28632	31209	29283	33648	35484	38205	44377	265496
R&A Hill Plymouth	22105	26717	31688	33719	36209	33366	41557	44827	270188
Aberdare Company		6512	13247	25022	22312	21458	29072	33881	151504
Blakemore Melingr'th				4157	7650	9755	11571	9953	43086
Brown, Lennox			2147	1836	2014	2810	7047	9907	25761
Gadly's Iron Co.				414	559	945	5400	3553	10871
Taff Vale Iron Co.						7200	13962	14346	35508
Bute Co. Rhymney					166	1042	273		1481
Totals	123732	154321	183675	234862	235757	307004	367756	395420	2002527

Iron carried on the Monmouthshire Canal

dates

Works	*1803–6*	*1807–10*	*1811–4*	*1815–8*	*1819–22*	*1823–6*	*1827–30*	*Totals*
Blaenavon	24425	35307	52956	45626	31003	39112	37542	265971
Garnddyrys				7344	13539	15250	10987	47120
Beaufort	13156	15465	14255	11135	12004	22735	23576	112326
Clydach	4767	4667	6292	13216	15286	14676	23028	81932
Ebbw Vale	8809	7306	17972	13505	17513	39363	64974	169442
Varteg	4428	8853	844	127	8795	19961	33778	76786
Tredegar	4080	26468	31233	35332	33376	43321	53830	227640
Nant-y-Glo			5392	25411	38126	55905	71639	196473
Coalbrooke Vale					2799	9593	10579	22971
Blaina						3898	15556	19454
Pentwyn						821	18410	19231
Abersychan							21966	21966
Bute							21957	21957
Others	15913	13315	25195	18823	20452	26859	23892	144449
Totals	75578	111381	154139	170519	192893	291494	431714	1427718

Tonnage of iron carried on the Monmouthshire canal & tramroads

Works		*1831–2*	*1833–4*	*1835–6*	*1837–8*	*1839–40*	*Totals*
J&C Bailey	Nant-y-glo	39199	43601	50341	49244	51607	233992
Kendall	Beaufort	11202	2257				13459
Harfords	Ebbw Vale–Sirhowy	38518	39454	48512	46054	49541	222079
Kendricks	Varteg	22690	25389	26040	7753		81872
Varteg Iron					12526	25489	38015
British Iron	Abersychan	15773	15725	22002	22118	24771	100389
Hills & Wheeley	Blaenavon	18692	16688	16629	5058		57067
Hills & Wheeley	Garnddyrys	5966	5814	8543	4159		24482
Frere	Clydach	6231					6231
Clydach Iron		6542	13513	15258	16364	19644	71321
S. Homfray	Tredegar	26644	25181	26042	28167	30149	136183
Coalbrooke Vale		4585	5327	6965	7325	13028	37230
Russell	Blaina	10881	11204	18144	14157	1140	55526
Hunt Bros.	Pentwyn	8888	8463	14734	8743		40828
Pentwyn Iron					2614		2614
Bute	Rhymney	14728	9273	7444			31445
C.H. Leigh	Pontypool	5390	8141	8594	14404	17586	54115
Daniels	Abercarn	1912	279	2147	1759	1060	7157
Pontymister		2411	466	2218	3359	2413	10867
W. Russell	Machin	1366	363	2053	1584	961	6327
J. Brown	Tydee		15	151	274	71	511
R. Walker	Cwm Tylery	64	66	1044	266	217	1657
J. Kennedy	Lanhiddel		25	258	727	174	1184
J. Jenkins	Pontypool	1117					1117
J&C Bailey	Beaufort		15063	27543	22048	20554	85208
Blaenavon	Blaenavon				10166	13065	23231
Blaenavon	Garnddyrys				7144	13663	20807
Golynos					3536		3536
Pentwyn & Golynos					2982	30316	33298
Rhymney Iron				2963	20344	33462	56769
Cwm Celyn & Blaina Iron						13600	13600
Others		573	168		769	8496	10006
Totals		243372	246475	307625	313644	371007	1482123

Acknowledgements: Diolch

We owe a great debt to many people; a few only can be shown, omission is not intended to discredit.

Geoff Smith of the Monmouthshire Railway Society who gave us permission to use his notes which first appeared in the Monmouthshire Railway Society's Journal and for the loan of many elderly maps. The late W.M. Porter, Manager, and the late Will Evans, foreman, of the then Neath Canal Navigation, together with Colin Powell, the Canals Officer of the Neath Canal Joint Management Committee. J.M. Davies, M.A. of Swansea and M.J. Bevan of Risca gave us much assistance with the first edition. Aldo Rabbaiotti of Baglan, Miss Johanne Brown of the Torfaen Museum Trust of Pontypool and Malcolm Morse of the Monmouthshire, Brecon & Abergavenny Canals Trust Ltd, gave much of their time and help in our 1988/89 survey. Martyn Denney, L.A. (Teddy) Edwards and Sheila Doeg, Editor of *Waterways News*, gave their unstinting assistance. Our publishers, in particular Jane Kennedy, unearthed material of considerable use and Kevin Lane, photographer-cum-author sorted out some of our pig's ears of negatives.

In 1974 we stated 'Gwell gwadd ym Merthyr na ym Manceinion!' True then and truer now.

References

1. Phillips, D.R., *The History of the Vale of Neath*, 1925.
2. Lloyd, J., *The Early History of the Old South Wales Iron Works (1760–1840)*, 1906.
3. Priestley, J., *Navigable Rivers & Canals*, 1831.
4. Lewis, S., *A Topographical Dictionary of Wales (3rd Edn.)*, 1845.
5. Coxe, W., *An Historical Tour of Monmouthshire Pt. I*, 1801.
6. Davies, W., *General View of the Agriculture & Domestic Economy of South Wales, Vols. I & II.*, 1815.
7. B&AC Committee Minute Book, 7th February, 1812.
8. B&AC Proprietors' Minute Book, 25th April, 1833.
9. Phillips, J., *General History of Inland Navigation*, 1792.
10. De Salis, *Bradshaw's Canals & Navigable Rivers*, 1904.
11. Rutland, Duke of, *Journal of a Tour through North and South Wales*, 1805.
12. TVR Proprietors Minute Book, 12th October, 1835.
13. IWA Bulletin No. 50, 1956.
14. Dupin, Baron, *The Commercial Power of Great Britain*, 1825.
15. Edwards, L.A., *Inland Waterways of Great Britain*, 1950.
16. Rees, A., *Cyclopedia Vol. VI*, 1819.
17. Jones, W.H., *History of the Port of Swansea*, 1922.
18. Oldisworth, Rev. J., *Guide to Swansea*, 1802.
19. Phillips, E., *A History of the Pioneers of the Welsh Coalfield*, 1925.
20. Overton, G., *A Description of the Faults or Dykes of the Mineral Basin of South Wales*, 1825.
21. Cumming, T.G., *Origin & Progress of Rail and Tram Roads*, 1824.
22. Wilkins, C., *The History of the Iron, Steel, Tinplate and Other Trades of Wales*, 1903.
23. *The Cambrian*, 12th April, 1806.
24. Hadfield, C., *The Canals of South Wales and The Border*, 1960.
25. *Abergavenny Chronicle*, 27th September, 1973.

Index

Aberdare
  Act of Parliament 9
  Dividends 11
  Extant remains 86
  History 5, 7–9, 11, 22
  Tolls 9
Aberdare Ironworks 9
Aqueducts *see* Engineering

Beaufort, Duke of 11, 42, 43
Blakemore, Richard 20
Boats 39, 42, 48, 53
Brecknock & Abergavenny
  Act of Parliament 11, 14
  Amenity usage 12, 87
  Extant remains 85, 87
  Dividends 11, 16
  Goods carried 16
  History 5
  Report by Outram 13
  Tolls 16
Burry Port *see* Pembrey Harbour
Bute, Marquis of 19, 21
Britton, *see* Glan-y-wern

Cardiff Railway 22
Clydach Ironworks 13
Cnel Bach 51
Crumlin Bog 22
Crymlyn & Red Jacket *see* Glan-y-wern
Cyfarthfa Castle 87
Cyfarthfa Tub-boat 52

Dadford, T. Jnr. 7, 9, 34
Dafen Pill 51
Davies, Elizabeth 24
Doctors 52

Earl of Ashburnham's 52
Engineering
  Aqueducts 42, 45, 87, 91, 99
  Breaches 19, 42
  Inclined Planes, canal 53
  Inclined Planes, tramroad 57, 71, 77, 80
  Locks 16, 20
  Reservoirs 20
  Tunnels, canal 16
  Tunnels, tramroad 60, 82
Englishman's Ditch 51

Giants Grave 34, 99
Glamorganshire
  Act of Parliament 19
  Bute's opposition 19–22
  Dividends 16, 21
  Engineering features 15, 17, 18, 91
  Extension to Lower Layer 19
  Extant remains 89
  History 5
  Opening ceremony 16
  Restoration scheme 22
  Water supply 20
Glan-y-wern 22, 23, 45
Grand Towy 53
Great Western Railway 29, 34, 43, 45, 77, 89

Hill, Richard 20

Inland Waterways Association 89

Jenkins, Richard 24

Kidwelly & Llanelly
  Act of Parliament 53
  Boats 53
  Dividends 53
  Inclined Planes 53
  Tolls 53
Kirkhouse, William 7, 45
Kymers 52

Llansamlet *see* Smiths
Llanwern *see* Glan-y-wern
London & North Western Railway 34

Melingriffith Works 87
Merthyr, Tredegar & Abergavenny Railway 78
Midland Railway 45
Monmouthshire
  Acts of Parliament 24
  Dividends 29
  Extant remains 23, 89, 96
  Extension to Newport 24
  Fourteen Locks Scheme 89–99, 106
  Goods carried 25–27, 109
  History 24
  Restoration schemes 89–99
  Tolls 29
Monmouthshire Railway & Canal Co. 29, 34
Neath
  Act of Parliament 34
  Dividends 39
  Extant remains 99
  Extension to Giants Grave 34
  History 34, 39, 105
  Tolls 34
  Water supply to factories 99
Neath & Swansea Junction *see* Red Jacket
Newport, Hereford & Abergavenny Railway 78
Outram, Benjamin
  Reports 13, 26, 56

Pembrey Harbour 53
Pen-Clawdd 51

Red Jacket 45
Risca, Magor & St Melons Canal Preservation Society 89

Sheasby, Thomas 7, 19, 34, 42
Smith's
  Extant remains 99
  History 39
South Wales Railway 53
Swansea
  Act of Parliament 39
  Breach of 1805 42
  Dividends 43
  Extant remains 99–101
  Goods carried 42
  History 5, 39
  Tolls 42, 43
  Water supply 42
Swansea Vale Railway 43

Taff Vale Railway 21
Tennant
  Boats 48
  Extant remains 101
  Goods carried 48
  History 45–48
  Opening ceremony 48
  Water supply to factories 101
Tennant, George 45
Tolls
  Aberdare Canal 9
  Brecon & Abergavenny Canal 16
  Hereford Railway 60
  Kidwelly & Llanelly Canal 53
  Monmouthshire Canal 29
  Neath Canal 34
  Penclawdd Canal 51
  Swansea Canal 42–43
Townshend, Chauncey 39
Tramroad exploration
  Craig-y-Hafod 82–84
  Grosmont 78
  Hereford 76, 78
Hills 79, 80
  Llanfihangel 78
  Llangattock 77

Tramroads
  Abercarn 65
  Abernant 57, 71
  Aberpergwm 71
  Abersychan 65
  Baileys 57
  Beaufort 57
  Blaenavon 55, 56, 65
  Blaendare 65
  Brecon Forest 73
  Brynoer 57, 58
  Caerleon 65
  Castle Morlais 63
  Claypонds 73
  Clydach 73
  Clyne 73
  Cribarth 73
  Cwm-mawr 65
  Dinas Uchaf 63
  Doctor Bevans 71
  Dowlais 63
  Ebbw Vale 65
  Graig-yr-Hafod
    Extant remains 66
    History 60
  Grosmont
    Extant remains 78
    History 60
  Hafod 63
  Hay 60
  Hen-Neuadd 73
  Hereford
    Extant remains 78
    History 60
  Hills
    Construction of 57, 62, 69
    Extant remains 80
    History 60
  Llanhilleth 65
  Llammarch 60
  Llanfihangel
    Extant remains 78
    History 60
  Llangattock
    Extant remains 72, 77
    History 60
  Llwydcoed 57
  Maes-mawr 63
  Nant-y-Glo 65
  New Lodge 65
  Nine Mile Point 68
  Old Waggonway 73
  Oystermouth 73
  Palleg 73
  Park Mile 65
  Parsons Folly 73
  Penderyn 57
  Pen-llwyn 65
  Pentrych 63
  Penydarren 56, 63, 64
  Pont Nedd Fechin 73
  Pontypool 68
  Pwll-Du 60
  Porthmawr 68
  Powells 63
  Rassa 68
  Rumney 68
  Scott 73
  Sirhowy 68, 69
  Tredegar 68
  Trevil 68
  Trimsaran 65
  Trosnant 68
  Tyla Quarry 60
  Varteg 66, 67, 68
  Waun-y-coed 73
Tredegar Ironworks 26
Trewyddfa 42
Trowman's Hole 22

Waun Lwyd 72
Welsh Midland Railway 43
West Bute Dock 21

Yspitty 51